In *Intersexualization*, Lena Eckert persuasively focuses our attention on the epistemic logic and discursive practices that have worked to define the "heterorelational sex-gender-sexuality system" and to deal with those whose bodies and identities fall outside of its norms. Deftly combining analysis of practices in the psycho-medical sciences (which she names The Clinic) with those of cross-cultural anthropology (or The Colony, as she calls it), and drawing on an impressive range of case studies, Eckert opens up space for a multiplicity of narratives, identities, and embodiments. As she writes, "sex is always complex," and this book illuminates what is at stake in how that complexity comes to matter for bodies rendered troubling and troublesome.

Ruth Holliday, *Professor of Gender and Culture, Director of Research,*
School of Sociology and Social Policy, University of Leeds, UK

Lena Eckert's book focuses on the cross-cultural construction of processes of intersexualization, exploring and combining knowledge production in medicine, psychology, sexology, anthropology and gender studies. Moreover, *Intersexualization* delves into the colonial archive to make sense of current practices which have far-reaching implications for the lives of people. A book that deserves a broad audience.

Gloria Wekker, *Professor Doctor, Emeritus, Department of Gender Studies,*
Faculty of the Humanities, Utrecht University, the Netherlands

I cannot recommend this book highly enough for its scholarship and its importance to contemporary intersex rights activism around the globe. More than merely taking cues from Foucault, Eckert develops a genealogy of both clinical practice and colonialist domination of bodies cast—as she shows the intersexualized to be—as degenerate, and in need of containing, disciplining, and controlling in the service of creating the powerful careers of the famous psychologists, psychoanalysts, physiologists, and sexologists who had everything to gain from the biomedicalization of intersex in the twentieth century. The book furthers critical intersex studies to more effectively challenge the racist, classist, and heterosexist legacies of twentieth-century practices and the long shadow they cast over contemporary concerns with the sites wherever embodiment, human rights, and sexualities meet.

Morgan Holmes, *Professor, Wilfrid Laurier University, Canada*

Intersexualization

Since the 1970s, research into 'Intersex' has been a central fascination for feminist theorists seeking to make arguments about how men and women are created as social/gender categories. *Intersexualization: The Clinic and the Colony* takes the case of Olympic runner Caster Semenya as a starting point to explore the issue of determining sex, and the ways in which intersexuality is a 'threat' to the distinction between men/women, homosexuality/heterosexuality and white/black.

By focusing on the 1950s and the 40 years after, Eckert shows how what she calls intersexualization began in psycho-medical research at the Johns Hopkins Hospital in Baltimore and UCLA, and has from there spread into cross-cultural anthropological accounts conducted in Papua New Guinea and the Dominican Republic. With cross-cultural intersexualization having been largely neglected in recent literature on intersex, this timely volume describes how such intersexualization derives from the combination of medicalization and pathologization through two crucial parts. The first part, 'The Clinic,' describes historical psycho-medical material engaging with hermaphroditism ranging from Greek Mythology up to today. This is followed by 'The Colony,' which analyzes, in several close-readings, cross-cultural anthropological, sexological and psychoanalytical accounts contributing to cross-cultural intersexualization.

Enclosing a wide range of inter- and transdisciplinary approaches to heteronormative and dichotomously organized frames of knowledge and organization, this volume is essential reading for upper-undergraduate and post-graduate students within the fields of gender studies, social studies of medicine, anthropology, science and technology studies, cultural studies, sociology, and history of medicine.

Lena Eckert is Assistant Professor in the Media Faculty at Bauhaus-University Weimar, Germany.

Routledge Advances in Critical Diversities
Series Editors: Yvette Taylor and Sally Hines

Intersexualization

The Clinic and the Colony

Lena Eckert

LONDON AND NEW YORK

First published 2017
by Routledge

2 Park Square, Milton Park, Abingdon, Oxfordshire OX14 4RN
52 Vanderbilt Avenue, New York, NY 10017

Routledge is an imprint of the Taylor & Francis Group, an informa business

First issued in paperback 2019

British Library Cataloguing in Publication Data
A catalogue record for this book is available from the British Library

Library of Congress Cataloging in Publication Data
A catalog record for this book has been requested

ISBN: 978-1-138-19330-7 (hbk)
ISBN: 978-0-367-34818-2 (pbk)

Typeset in Times New Roman
by Wearset Ltd, Boldon, Tyne and Wear

Für meine Eltern

Contents

Acknowledgments

The time I spent writing this book I used to live and work in three different countries: England, the Netherlands, and Germany. The many different people whom I met during these years have influenced and coined this book—they are too numerous to mention. Yet, some of them need to be named and thanked: First of all I want to thank my mentors who have influenced this book with their knowledge and their impressive personalities: Gloria Wekker, Rosemarie Buikema, Nancy Harding, and Ruth Holliday—you all have helped and supported me immensely, I am very grateful for discussions and critique. Moreover, thanks to Robert Davidson, Zowie Davy, Morgan Holmes, Christian Scholl, Eliza Steinbock—our exchange, and our discussions as well as your comments on the ideas for this book have helped me develop my arguments and understand the academic responsibility that comes with writing about intersexualization.

I would also like to acknowledge the kind assistance of Annabel van Baren and Anna Wegricht for help with the manuscript. Financial support for this project was provided by the Research Institute for History and Culture at Utrecht University, NL and the Centre for Interdisciplinary Gender Studies at Leeds University, UK. A generous grant by the equal opportunity office of Bauhaus-University Weimar in Germany has enabled this book to be published.

I want to thank my partner Georg Bosch, who knows what I feel for him besides gratitude for supporting me unconditionally and our amazing children Juri and Lukas who have made me understand what is really important in life. Finally gratitude goes to my brother and to my parents especially for believing in me and giving me all the support a child can wish for. Ich danke Euch! All of you have been part of the writing process of this book with your willingness to care for my project and me. Thank you!

Additional Acknowledgments

An earlier version of Chapter 6 was published as "From the 'Polymorphous Perverse' to Intersexualization. Intersections in Ethnographies," in *GJSS. Graduate Journal of Social Science, 10*(1) in 2013.

Introduction

"I am who I am and I'm proud of myself"

At the 2009 World Championships in Berlin, Germany, newcomer 18-year-old South African Caster Semenya won the gold medal in the 800-meter sprint. But it was her deep voice and flat chest—not her athletic success—that garnered the attention of sports officials, the popular press, and the public. Suspicions circulated that she was not really female and a worldwide controversy ensued over whether her assumed intersex*[1] body would offer an illegitimate advantage over her fellow female competitors. She became what I will call *intersexualized*. The International Association of Athletics Federations (IAAF) ordered Semenya to undergo gender testing after she had been awarded the gold medal. This gender- or sex-verification test requires numerous medical evaluations that involve gynecologists, endocrinologists, psychologists, and internists, as well as a gender expert.

Semenya excelled in the white domain of the 800 meters, and her case triggered widespread debate. I argue that Semenya's case—although purportedly exclusively about her sex_gender—also carried a racial dimension. Leonard Chuene, the head of South African athletics told *The Associated Press* that "if it was a white child, she would be sitting somewhere with a psychologist, but this is an African child" and he added: "who are white people to question the makeup of an African girl?" (*The Observer*, UK on August 23, 2009) The ensuing debate in the media all over the world included statements about the sexist, racist, and neo_colonial aspects concerning Semenya's treatment in the proceedings of the IAAF. Various scholars have since interrogated the panoptical gaze of the media, which Semenya was subsequently subjected to and the gendered and racial panic that erupted around Semenya was discussed widely (e.g., Cooky, Dworkin, & Swarr, 2013; Nyong'o 2009; Watson, Hillsburg, & Chambers, 2014; Vannini and Fornssler 2011). In professional sport, notions of race, gender, sexuality, nation, ability, and more, are being negotiated, shaped, and reshaped. Because sport inhabits a highly visible position in public media it is a domain in which public spectacles are portrayed and which thereby reflects on issues relevant to society at large. The case of Semenya covers a number of aspects in, what I call *cross-cultural intersexualization*; this book enfolds the many aspects present in the quest for a scientifically verifiable distinction between the sexes genders and in the accompanying racializing processes.

The athletic body is often relied upon to support the claim that sex is a natural fact: men run faster, throw further, and are generally stronger than women. When controversies such as Semenya's arise, which trace back as far as the 1900s when women started to participate in sports competitions, the issue is a question of the sex of the participant (Karkazis, Jordan-Young, Davis, & Camporesi, 2012). It follows then, that the battery of tests administered to the body in question would be referred to as sex verification. Yet the performed test is known as a gender verification test. Sex and gender are used interchangeably in a wide variety of circumstances and we need to ask why. The answer matters because in intersexualization it becomes clear that "sex is not a biological property, but rather a site of cultural negotiation" (Brady, 2011: 8). Vannini and Fornssler (2011) state that "there is no consensus with regard to these linguistics categories any more than there is consensus regarding the biological underpinnings of what constitutes sex" (249). However, gender-verification tests "constitute one element in a matrix of surveillance and policing practices of the boundaries around gendered bodies" (Cole, 1994: 20). The validity of gender-verification testing was never questioned in the debate.

Semenya was tested to verify—according to sex determinants—that she was and could participate as a woman. The World Championships, like the larger social order, had no place for her as intersex* because intersexualized people challenge reigning social, political, and bio-medical assumptions about the congruence between sex and gender. Like most athletic competitions, the World Championship is divided into two separate fields: women and men. It is expected that the men are 100 percent man and the women are 100 percent woman. Despite the wide variety between biological men on the one hand, and biological women on the other, regarding size, weight, strength, muscle density, hormonal levels and so on, this division is upheld. The myth of sports as a level-playing field however also reproduces other forms of inequality. As in the case of Semenya, it becomes clear that only certain bodies are privileged in international sporting contexts. Non-privileged bodies are subject to racism, classism, ableism, and other forms of discrimination. "Maybe it's because we come from a disadvantaged area," Semenya's coach Jeremiah Mokaba said "they couldn't believe in us" (Levy, 2009). The Olympics, the World Championship, and sports competitions in general are the emblematic institution of a racialized gender-divide based on sex, as this case proved.

I begin with the story of Caster Semenya because what seems to matter most in this case is less the possibility that she is a man and not a woman, it is the possibility that she is *neither* a man *nor* a woman. And the procedure in which Semenya was discussed as possibly intersex* testifies to the ways in which race and neo_colonial legacies intersect with the problem of sex and gender. Her intersexualization was a cross-cultural process that ignored local understandings of sex and gender in favor of Western scientific classifications. The South African newspaper *Sowetan* published an article called "The IAAF is a Disgrace" claiming that "the conduct of the international body was racist and humiliating" (quoted in Cooky et al., 2013: 40). *Sowetan* furthermore noted, "we all know that her crime is that an African girl outran everybody" (Mofokeng, 2009). As sports sociologist Cheryl Cooky and her colleagues state, for the Global South, "Western scientific classifications of raced and gendered bodies are viewed as products of colonialism, European expansionism, and racism" (2013: 34).

Cross-cultural Intersexualization

In the case of Caster Semenya, the agenda to intersexualize came from somewhere *else*. The tests were not conducted in her homeland of South Africa but in the IAAF host country Germany. Whatever Semenya's body would reveal, it would be interpreted in the Western framework of binary sex determination. Semenya's intersexualized body emerged through its production as other-than-the-binary-sexed-body and as other-than-the-Western-body. According to Nirmal Puwar (2004), the bodies of women and/or racialized minorities become seen as 'space invaders' when moving into white male domains. Semenya can therefore be described as a "sporting space invader" (Brown, 2015). In his article on *The Spectacle of the Other* Stuart Hall (1997) shows how not just Black sexuality but also the Black sporting body has been produced throughout neo_colonial history as a marker of the supposed primitivism of the Black body. He demonstrates how especially female Black bodies are constituted as other via their transgression of first the boundaries between male and female and second between human and ape. For example, Hall quotes a 1988 *Sunday Times Olympic Special* that featured photos of a Black female American athlete. Her husband and her brother said here: "Someone Says My Wife Looked Like A Man" and "Somebody Says My Sister Looked Like A Gorilla" (232). The non-normative gender performance of the female Black sporting body is made to signify the neo_colonizing gaze. The representations of Black bodies are inextricably inscribed with the fantasies and anxieties of racist and neo_colonial histories. They are inscribed with a hierarchy of exhibition that positions Black bodies at the very margins of the human. The simultaneous display of the 'uppity' of African-American sportswomen and their bodies as sexually grotesque and pornographically erotic has been discussed (McKay & Johnson, 2008: 492). As Letisha Cardoso Brown argues, it is not just visual representations, but all kinds of "narratives of black female bodies in the west" that are tied to the legacies of slavery and "of colonialism, apartheid and other forms of oppression" (2015: 16). Black women are often depicted in "sexualized," "lascivious," "wild," "primitive," "animal-like," "unfeminine," "welfare queens," or "matriarchal" frames (Cooky, Wachs, Messner, & Dworkin, 2010: 154). These modes of portrayal have historically operated to 'other' women of Color. This is of course also due to the shaping of hegemonic standards of beauty and femininity by whiteness and white privilege.

The *Los Angeles Times* quotes South Africa's ruling ANC party and the Young Communist League of South Africa as having stated that "it feeds into the commercial stereotypes of how a woman should look, their facial and physical appearance, as perpetuated by backward Eurocentric definition of beauty" (Dixon, 2009). Moreover, as Cooky et al. state, in South Africa:

> there is a historically specific understanding of sex testing as linked to and identified with Western scientific classifications of gendered and raced bodies. In this manner, sex testing aligns with past colonialist exploitation and contemporary forms of racial oppression, even in the postapartheid context.
>
> (2013: 33)

And they continue, that "for the Global South, Western scientific classifications of raced and gendered bodies are viewed as products of colonialism, European expansionism, and racism, not simply 'objective' or 'value-free' accounts that ensure equality in sport or in South African society" (33). South African voices that were reported, when Semenya turned home sounded like this: "it is the ghoulish, white-coated scientists of the IAAF who would do well to look into their hearts and ask whether the overwhelming evidence of Caster's life as a girl in South Africa does not count as science" (Brooks, 2009). Leonard Chuene, the head of South African athletics, pronounced the treatment of Semenya as "racism, pure and simple" and asked, "who are white people to question the makeup of an African girl?" and he continues: "In Africa, as in any other country, parents look at new babies and can see straight away whether to raise them as a boy or a girl. We are now being told that it is not so simple" (Smith, 2009a). Chuene here dismantles Western science as being neo_colonial and racist. In the history of professional sports a number of incidents like this one have happened. The reason Semenya's case was so controversial is that she is South African and that Euro-American institutions conduct all these tests. This fact interlinks the dimensions of sexism and racism in a neo_colonial geo-political setting.

Semenya and Baartmann

Semenya's story calls to my mind the story of Saartje Baartman, the Khoikhoi woman removed from South Africa and displayed naked throughout Europe due to her unusually large genitals and protruding buttocks in the nineteenth century. Physical examinations of Baartman who circulated in European discourse as 'Hottentot Venus,' focused on her body as "an essentially different type of the human species" (Buikema, 2009: 75). The analogy of her genitals to those of animals created her as an inferior race or species to the white civilized Europe-ans, who paid money to examine, dissect, and exploit her. Baartman's sexual anatomy signified to both scientific experts and average citizens the kind of primitive, savage, excessive, uncivilized, and untamed nature attributed to non-European bodies and cultures at the time, in part to justify colonial expansion. Semenya's twenty-first-century story is not so very different. Although not dis-played naked for the world to see, Semenya's intersexualized body has also been exploited as a spectacle and has been displayed for public consumption. In 2010 the search phrase 'Caster Semenya' in Google revealed eight million hits.

The discussions of the cases of Semenya and Baartman are characterized by the opposite construction of sexuality in the other: excessive masculinity and femininity respectively. Yet, their representations work in a similar fashion. The objectification of both reduces them to their sexuality and to their race. The cultural archive that draws upon these specific sets of sexualized and racialized representations, deriving from the imperialist and colonial history, proves to be alive and kicking. Baartman was othered in a time in which colonial endeavors and violence required justification through the inferiority and animality of the colonized. In the globalized and neo_colonial age, Semenya is othered through intersexualization, which is associated with developmental issues (DSD) and ambiguity of sex-gender. This happens at a time when the sexed_gendered basis of Western civilization and consumption, the division of production and reproduction threaten to dissolve and cause anxiety. Representations of non-Western sexuality in both cases serve to affirm and reinstall the achievements of Western civilization and its sex-gender ideologies according to the justifications needed in a particular time.

Baartman was a rare, exotic, colonial curiosity; Semenya was the subject of the normalizing and medicalizing processes emblematic of contemporary times. Semenya appeared on the cover of a magazine with a complete feminine make-over shortly after the media hype about her suspected maleness began. This cover photograph is an image of Euro-American femininity: make-up, painted fingernails, a dress, jewelry, and a feminine posture. The title screams, "Wow, look at Caster now!" and is the headline on South African glossy magazine *YOU* (Number 114, September 10, 2009). The photograph is presumably to convince everybody that Semenya is a woman—to defeat the rumors that she might be a man or at least that she is too masculine. Semenya and Baartman are connected by the pathologization of their imagined difference from the register of norms imagined in the discourse of Western science—albeit in opposite ways.

If we compare Semenya's case to representations of Baartman, significant differences emerge that show that Baartman was represented as excessively female, animal-like and primitive. Semenya on the other hand was represented as excessively masculine and monstrous in her lack of femininity. The *YOU* magazine cover picture was meant to displace this. As Vannini and Fornssler aptly analyze, Semenya does 'lack' these same feminine qualities and it follows then that she must be a man. In short, she "cannot be a woman because she is lacking-the-lack" (2011: 250). Both, Semenya and Baartman are produced by a Western gaze that claims the power to determine, name, label, and assign all bodies a place in the (always) racialized Western sex-gender-sexuality-system. In both cases, Western scientific institutions dissect, display, diagnose, and exhibit. The spectacle that Baartman was subjected to seems to have been repeated. The freak show continues. Despite the century between the two cases, there is remarkable continuity in the neo_colonial that produces African women as inadequately sex(ualiz)ed.

It is not just that Western scientific registers are called upon to determine if Caster Semenya can participate in global sport competitions; Semenya's own voice was hardly heard in the process. "The exclusion of Semenya's voice and her subsequent invisibility deflected any challenge that her perceptions of her own experience would pose to the 'controlling images' of Black women" (Cooky et al., 2013: 47). Semenya's agency was constrained completely, not just by Western media but also by South African political leaders and stakeholders who discouraged her from raising her voice during press conferences. "Indeed" as Cooky et al. report, "Semenya was quoted in only 5 (approximately 9 percent) of United States newspapers and in only 8 (approximately 5 percent) of South African newspapers" (2013: 46). What did not matter was Semenya's own perception of herself. Her expertise on her own body was ignored and she was discouraged to speak up. The IAAF banned her from competitions while completing the investigations. However, in 2010, Semenya was cleared to compete by the IAAF and her Berlin victory was allowed to stand. She participated in both the 2011 World Championships and the 2012 Olympic Games in London.

Introducing Intersexualization

Semenya's case allows me to introduce the range of issues I cover in this book: first intersexualization is based on the distinction between sex and gender, i.e., the gender-concept. Second, the negotiation of the distinction between sex and gender and between male and female results in the casting of bodies that do not fit the current parameters of sex-gender determination as intersex*. Third, these 'unruly' bodies are medically diagnosed even though they are not sick, and surgically modified; their being inter* is thus erased. Fourth, the multiple discourses, which participate in intersexualization are drawn from psychoanalysis, anthropology, sexology, bio-medicine. Fifth these are intrinsically intertwined with neo_colonial and racializing modes of knowledge production which cause cross-cultural intersexualization.

Judith Butler also reflected on Semenya's case and stated, that if the panel for the 'gender'-verification test included:

> a gynaecologist, an endocrinologist, a psychologist and an expert on gender, then the assumption is that cultural and psychological factors are part of sex-determination, and that no one of these "experts" could come up with a definitive finding on his or her own.
>
> (Quoted in Brady, 2011: 8)

She suggests, that if the result of the tests has to be reached by consensus between these experts, then if there is no consensus, there cannot be a sex— concluding that the IAAF acknowledges that this "simultaneously demonstrates that the determination of gender does not take place on a discursively neutral terrain" (Ibid.: 9). The acknowledgment of the socio-political nature of this discussion does however not have any consequences, neither for sex testing, nor for the gender separation in professional sports. What did change is, that in 2011 the IAAF released a new policy on the procedures of determining the eligibility of female athletes with hyperandrogenism. It states that they may be eligible to compete if they submit to having their androgen levels tested. If these are higher than the "normal male range" or if they are within the "normal male range" deemed to derive a competitive advantage they could be ineligible for competing as a woman. This gives priority to hormones—here androgens—in the determination of sex. It also continues to frame androgens as markers of ability (Karkazis et al., 2012). Sex testing, as Karkazis et al. state, continues to be "problematic because there is no single physiological or biological marker that allows for the simple categorization of people as male or female" (6). In most cases of sex testing, then one category has to be given prevalence in the determination. One factor has to be singled out to be strong enough to enable a final decision. Here, in the case of hormones taken as the final arbiter, Karkazis et al. conclude that "there is a great deal of mythology about the physical effects of testosterone and other androgens" (8). Sex or gender testing is therefore a complex procedure based on various, historically changing parameters, and negotiation between 'experts' from different disciplines and apparently also a good part mythology.

Not just in professional sports but in society in general, intersexualization, that is the drawing of the boundaries between binary sex and gender produces bodies that are abject and cannot belong to either category. The medicalization and pathologization that went hand in hand with these processes had and still have horrendous consequences for the 'bodies in question.' They are subjected to immense scrutiny, medical treatment, and surgery. It is only recently that these surgeries and medical treatment paradigms have been questioned. This debate of the last 30 years has now achieved wide reaching attention—also outside of the medical realm and has entered sociological, psychological, cultural, juridical, human rights, and historical attention. My contribution to this debate however, extends the focus that has been placed upon intersexualization in the Global North to research conducted in the Global South—by researchers coming from the United States of America. With this I bring the problem of racialization and neo_colonial legacies in research into sex and gender to the center of attention in cross-cultural intersexualization.

Nomenclatures, Timeframes, and Terminology

In twenty-first-century medical nomenclature, Semenya's body is no longer 'hermaphroditic'—a term that was slowly replaced with 'intersexual' starting in the 1950s. Detailed bio-medical taxonomies of various 'intersex conditions' are a precondition for the 'intersex' diagnosis to be made. Alice Dreger and April Herndon (2009) refer to "variations in congenital sex anatomy that are considered atypical for females or males" and conclude that "the definition of intersex is thus context specific ... so the definition of intersex depends on the state of scientific knowledge as well as general cultural beliefs about sex" (200). Intersexualization is thus a historically variable process which, depending on time, space, and so-called experts changes the definitions of who is male, female, or inter*.

Critical feminist biologist Anne Fausto-Sterling lists six so-called 'common intersex' types, the definitions of which can vary somewhat: Androgen insensitivity syndrome (AIS); Congenital Adrenal Hyperplasia (CAH); Gonadal Dysgenesis; Hypospadias; Turner Syndrome; and Klinefelter syndrome (Fausto-Sterling, 2000). Even though for an understanding of the issue of intersexuality, the repetition, contextualization, and framing of medical diagnosis is necessary, such lists reflect the medical establishment's compulsion to diagnose and repeat the violent gestures of intersexualization. As I have argued elsewhere, intersexuality could be understood as an "identity based on the experience of medical treatment" (Eckert, 2009: 41). In this book, I will argue that there are several other aspects complicit in the processes of intersexualization; one of them is the medical naming and framing that precedes the treatment. As described in *Paradoxes of Gender* by Judith Lorber (1994), one is equally caught in a paradox when talking and writing about intersexuality, just the same as in the case of gender. In order to remain intelligible, one needs to repeat the medical taxonomies even though they will be reinstalled as legitimate and thus reproduced. In this book, I am caught in this paradox. I will however try to escape certain ascriptions and will use inverted commas or underscores whenever I see it fitting.

This book represents my contribution to the growing body of work examining the quest for a scientifically verifiable distinction between the sexes that traces back to the ancient Greeks, as Laqueur details in *Making Sex: Body and Gender from the Greeks to Freud* (1990). The time frame of my investigation into intersexualization begins more or less with Freud and contemporary emerging discourses such as sexology. I examine how, from the moment intersexuality appeared center stage in mid-century American psycho-medical publications as a phenomenon that would seemingly question the 'fact' that femaleness and maleness stand in contradiction to one another, it was in fact quickly absorbed into the prevailing paradigm that posited femaleness/femininity and maleness/masculinity as mutually exclusive, binary opposites. It is often assumed, as Karkazis et al. note, that the bodies of intersexualized people are somehow exceptional because of their complexity, but sex is *always* complex. Biologists know: there are many biological markers of sex but none can be ranked decisive. Not one marker of sex "is actually present in *all* people labeled male or female" (2012: 6). The assumed binary between male and female constantly needs to be re-mediated—according to new psycho-medical findings—via the pathologization of some individuals that are made to not fit.

I use the term hermaphroditism or hermaphrodite whenever it is used in the body of research that I am analyzing. I do not want to indicate that it is not a constructed medical category; however for better readability I will not put it in inverted commas. I use the term of 'heterorelational sex-gender-sexuality-system' to indicate that the categories of sex, gender, and sexuality are intrinsically connected in the gender-concept. Sometimes, however I will replace it by the term 'heterorelational system' for better readability. I will use the term sex(ualiz)ed_gendered to indicate the processes of intersexualization that have brought about the distinction between sex and gender. The underscore _ is used in multiple words and indicates the difficulty to separate some concepts in specific contexts. Whenever I use sexed_gendered I indicate that the two concepts cannot be clearly distinguished—there is no reason or justification for using either_or. The brackets around the '(ualiz)' in 'sex(ualiz)ed' indicate that the sexed body as the biological body is not to be separated from its sexualization, emphasizing the penetration by the heteronormative matrix which requires each and every body to have a sexual orientation/preference. The body is, according to the 'direction' of its desires, located along the coordinates of homo- and heterosexuality but also on the coordinates of normality/abnormality and male/masculine and female/feminine (Hark, 1993). Pathologization and normalization are intrinsic in this process and are based on the supposed naturalness of heterosexuality/reproduction. I use the term neo_colonial to indicate that everything in neo-colonial is not to be separated from the colonial. The underscore also presents itself in this book as s_he, him_her, and his_her to designate a space by the _ that includes people who feel that the pronouns or the pronominal adjectives of he/she, his/her, him/her do not denote their identities and to indicate a space that allows for greater diversity than gendered English grammar.[2]

Most of the work on hermaphroditism/intersexuality views it as a 'phenomenon.' On the contrary, I argue that it is the perception of the 'phenomenon' that is the real phenomenon. I want to introduce the term *intersexualization,* a slight but significant shift away from intersex/intersexed/intersexual/intersexuality. In the intersexualizing process, experts first determine the true sex of the infant and then assign what they call the 'best sex' in their treatment recommendations. *Intersexualization* is useful because it draws our attention to the active processes that take place to render certain so-called 'ambiguous' bodies intelligible. Further, it takes the focus away from factual discussions of what intersexuality actually is, and emphasizes instead the epistemic logic and discursive operations that have initiated, manifested, and re-articulated the processes of intersexualization time and time again in psycho-medical and anthropological discourses.

More recently, the diagnosis of Disorders of Sexual Development (DSD) has begun to replace intersexuality. At a 2005 conference in Chicago, 50 international medical experts formulated a consensus statement proposing the new term and defining it as "congenital conditions in which development of chromosomal, gonadal or anatomical sex is atypical" (Hughes, Houk, Ahmed, Lee, & Group, 2006: 554). DSD is an umbrella term for a variety of intersex conditions and has caused a great deal of discussion amongst activists, scholars, and physicians (see Feder & Karkazis, 2008; Holmes, 2008; Spurgas, 2009). Whereas DSD is currently the preferred term in medical circles, I am highly skeptical of supporting the narrative of abnormal development it suggests. DSD revives late nineteenth-century sexological discourses that emerged alongside a multiplicity of evolutionary discourses that permeated racial and sexual thinking and practices. As I will show throughout this book, the intersecting discourses of race and gender are complicit in producing cross-cultural intersexualization; the notion of development towards is hereby crucial.

According to Foucault, the gaze of psycho-medical discourse is a purposeful looking that does not merely perceive objects but actually creates meanings and their boundaries. Canadian intersex activist and scholar, Morgan Holmes describes what I call this aspect of intersexualizations as "the medical gaze that *creates* as much as, or more than, it *describes* pathology" (Holmes, 2008: 114). Inspired by Holmes, I therefore, argue that intersexualization is the result of the bio-medical creation of bodies as male or female. It is not the fact of male and female that produces a third—the intersexualized body—but the very processes of intersexualization that produce white hegemonic heteronormative maleness/masculinity and femaleness/femininity as natural and normal. Intersexualization, I suggest, is at the core of the process of the construction of a dichotomously sex(ualiz)ed_gendered society. The category of intersexuality repeatedly re-articulates the distinction between male/female, masculinity/ femininity, and homo/hetero. To apply this shift concretely, I would suggest that Semenya is not intersexual but rather that she *became* intersexualized through this lengthy and trying ordeal. In other words, there was no ontological intersex body prior to the discursive process of being tested and assigned a sex and all of the institutions, scientific fields, and organizations that attended that process. This is, of course, not to say that intersexuality does not exist, quite the contrary, with US-American intersex scholar and activist Georgiann Davis I believe that "intersex people exist all around the world" (2015: 3). In 2009, Iain Morland, intersex scholar and activist from the UK, demanded that "future research on intersex should continue to interrogate the multidisciplinary contexts in which its medical management has emerged" (195). This book tries to answer this request by analyzing the psychoanalytical, anthropological, and bio-medical legacies in what I call intersexualization.

The Clinic

This book is divided into two primary parts: The Clinic and the Colony. Part I on the Clinic focuses on a close-reading of research by US-American scholars who established themselves as 'experts' on intersexuality in order to discern the process of intersexualization inherent in these works. This part therefore mainly interrogates the rationales for claiming genital surgery is necessary for inter* infants since the 1950s. Researchers have intersexualized certain bodies, in the sense of suggesting to erase their state of being inter* with surgery and hormonal treatment, starting with the invention of a formal distinction between sex and gender. Key researchers at Johns Hopkins University and the University of California, Los Angeles construed intersexuality as a psychopathology in need of treatment, despite the fact that their own studies uncovered no actual psychological or medical problem with the patients in the study. They used intersexuality to develop concepts such as gender role (Chapter 1) and gender identity (Chapter 2). Based on selective readings of Freudian psychoanalysis, various arguments were developed that fostered intersexualization—misogynistic and homophobic inclinations are to be discerned in the process. Feminist theorists, however, in order to confront gender discrimination happily adopted the gender-concept developed in intersexualization. Yet, its implicit reference to dimorphic sex as a biological fact made the deconstruction of gender difficult and enduring. Feminist biologists such as Anne Fausto-Sterling have taken up the challenge and tackled intersexualization directly. The linguistic and conceptual limitations, which come with the gender-concept, however have made this enterprise difficult. New Materialist Feminists with their different conceptualization of sexual difference and matter might be able to provide a way out of intersexualization (Chapter 3).

The Colony

In Part II on the Colony I will engage with what I call cross-cultural intersexualization. Cross-cultural intersexualization is the mode of transferring intersexualization that occurs in the Clinic into the 'field' of cross-cultural anthropology. What links Part I to Part II of this book is the critical interrogation of the theoretical underpinnings of theories on sex, gender resulting in discursive and material practices causing intersexualization. The cases I will interrogate in the Colony are located in Papua New Guinea and the Dominican Republic. The problems I see with intersexualization—the constant reiteration of masculinity and femininity as exclusive subject positions on the basis and attempt of the surgical and hormonal construction of two exclusive body morphologies (male and female)—is now combined with racializing/ethnicizing processes. In the Colony we find situations that are already laden with unequal power structures—just as in the Clinic between doctors/ researchers and patients—but here in the anthropological field multiple power structures intersect. The participating subjects in cross-cultural intersexualization— researchers and researched—are caught in a net of a variety of different power positions. Very specific knowledge and concepts about desire and the body are brought upon people who might have a completely different conceptualization of both. Intersexualization, when combined with cross-cultural research becomes neo_colonial since not just knowledge about the body and desire is imposed upon the other (in the very process of research as well as later on in the publication) and the original perception of the sex_gender_sexuality of the researched become eradicated and racializing and neo_colonial assumptions and processes are intrinsic.

The Chapters

Chapter 1 examines the mid-1950s series of articles that John Money co-published with John and Joan Hampson at Johns Hopkins that contains their recommendation that intersex* infants undergo corrective surgery. I argue that the specifics of Money's own concept of gender role drove him to argue for surgical modification of infants born with 'ambiguous' genitals. Money's body of research was the single most influential factor for the treatment paradigm of intersexuality in the last 60–70 years in most countries of the Global North. Yet, the underlying bio-medical, psychological, and socio-cultural rationales and theories have only recently been examined (e.g., Downing, Morland, & Sullivan, 2014; Fausto-Sterling, 2000; Kessler, 1998; Klöppel, 2010). My contribution to this growing body of work is an interrogation of the psychoanalytical underpinnings in Money's research. Money and his colleagues practiced a selective reading of Freudian psychoanalysis that allowed them to conceptualize psychosexual development as based on sexual dimorphism and consequently argue for the naturalness of a gendered dichotomy and its surgical and hormonal construction in intersexualized individuals. These normalizing surgeries, particularly clitorectomy, became common in intersexualization. Money's treatment recommendations remain widely influential in guiding the doctors who determined the sex of their young patients. One can assume that very few intersex* individuals escaped the surgical knife; there are however, no statistics to be found (see also Dreger 1999). Even though intersex activists have formed movements to defeat this practice, physicians predominantly continue cutting healthy flesh for the sake of an esthetico-functional outcome.

Chapter 2 turns to the work of psychoanalyst Robert Stoller, who slightly modified Money's gender-concept to come up with the category of gender identity as distinct from gender role, which Money established. In Stoller's theory, dimorphic sex evolves as the precondition for a binary organized gender identity; gender identity as the sense of being either male or female becomes first essentialized by a postulated biological force that, supposedly, naturally produces masculine and feminine identities. Stoller developed the principle that one *is* a sex and *has* a gender identity. To formulate his claims about a core gender identity, and articulate a specific hermaphroditic core gender identity, Stoller also selectively uses Freudian psychoanalysis, just as Money did. He conceptualized the infant–parent relationship and the bodily ego in neo-Freudian terms and altered them to intersexualize and transsexualize. Stoller came to claiming that there is a hermaphroditic identity that is caused by a hermaphroditic body. In this case, he echoed Money's treatment protocol, suggesting early surgery in order to prevent this hermaphroditic gender identity from forming. The gender-concept—that is, quite simply, the assertion that we all have one gender as either male or female that is determined by dimorphic sex—is fundamental for the normalization and pathologization processes we find in intersexualization. In other words, without the coinage of the gender-concept, intersexualization would have probably been harder to establish as the congruence between sex and gender helped immensely to justify their surgical alignment.

Since the coinage of the term in the 1950s in psycho-medical research, 'gender' has never stood completely alone, as I show in Chapter 3. Important theoretical attempts to temporarily delink it from 'sex' for the purpose of analysis and intervention aside, invocations of gender are always also invocations of sex. Feminists have, however, found the gender-concept useful in challenging sexism. As Judith Butler already argued in *Gender Trouble* in 1990, the gender-concept will always be problematic because it always implies sex as its natural biological foundation. However, critical approaches to the body by feminist and queer theory since the 1980s deserve credit for powerful interventions in the production of heteronormativity and the supposedly natural dimorphic sex(ual) difference as well as the 'deviances' or 'disorders' which are produced to reaffirm the norm. They reveal the unnaturalness and arbitrariness of heterosexual normalcy and have disrupted the powerful links between bodies, identities, and sexuality. Feminist biologist Anne Fausto-Sterling's approach to intersexuality was groundbreaking in many regards, yet, it still depends upon sexual dimorphism, even if her goal is to challenge or dismantle it. In 1993, Fausto-Sterling provocatively proposed we think of five—not two—sexes, in her initial attempt to deconstruct bio-medical categorizations of sex and gender. Chapter 3 discusses both the critical potential and the problematic side of her biological approach to intersexualization. In the last parts of the chapter I introduce the reader shortly to some accounts by the new feminist materialists to show how an intervention in intersexualization could be framed differently—with regards to the materiality of the body and sexual difference.

Chapter 4 interrogates the mid-1980s clinical anthropological research of Gilbert Herdt and Robert Stoller, who co-conducted an interview with a Papua New Guinean Shaman. It was not the shaman's 'shamanism,' however, but rather his presumed hermaphroditism/intersexuality that was of interest to them. Whereas they claim diagnosis was not their intention, they nevertheless diagnosed the shaman Sakulambei as a hermaphrodite. Their account repeats the spectacle of Sarah Baartman as it exposes, discusses, and exploits the sexuality and the body of what strikes them as unusual. They pathologize his body because it does not fit into their familiar coordinates of the Western heterorelational sex-gender-sexuality-system. The hegemonic, Western, medical gaze they exert upon Sakulambei reiterates the power structures present in the Clinic between clinician and patient. Yet, in this neo_colonial setting the clinical in ethnography hosts a number of problematic dimensions—the issue of translation in anthropology, the problem of the translatability of understandings of identity/self/subjectivity, ethical questions of extracting psychoanalytical data in an ethnographic setting, projections of the researchers upon the researched, and the powers of the visible and processes of invisibilization and thereby silencing—all dominant features in intersexualization now going cross-cultural and shifting power structures even farther.

Chapter 5 continues to focus on clinical anthropological research. Here, I look at the work by the physician Julianne Imperato-McGinley in the Dominican Republic and on further research conducted by Gilbert Herdt—now in Papua New Guinea in 1988 with the endocrinologist Julian Davidson. In the Dominican Republic case, Imperato-McGinley's medical research team tries to consolidate the postulated biological force by Stoller. They use blood and urine samples to diagnose male pseudo-hermaphroditism and conclude on the influence of hormones on the development of gender identity. The very presence of anthropologists and colleagues changed the attitudes of the communities towards the individuals under investigation. As Imperato-McGinley reports, the villagers have formerly not been familiar with the 'condition' they identified, but now that they are, affected individuals are objects of ridicule and acquired the derogative term of *guevedoche*. It seems that the examined 'phenomenon' caused no distress prior to the interference of the Western research team. Only after the researchers invaded the village were intersexualized individuals ostracized. The researcher team created a social category of sexual deviation that had not previously existed in the community. Herdt and Davidson's project researching the Sambia in Papua New Guinea constructs its sample population as 'naïve.' They pathologize as effeminate those who do not conform to their diagnosis of male pseudo-hermaphroditism. Typical of almost all medical anthropological research, both studies explain their findings within the framework of the Western heterorelational system, and both impose Western notions of identity—in the cases interrogated here notions of masculinity. In this framework every body can have and express a gender, but only the one that their sex—as diagnosed by Western bio-medicine—determines. However, Herdt challenges the Dominican Republic case and argues that there is a three-sex code present as opposed to a two-sex-system in the West. In these cases another dimension of neo_colonization is identifiable; both studies have colonized the other cultures with their representation in the framework of the Western heterorelational system. The imposition of the Western notion of identity and the silencing of their symbolic organization testifies to the neo_colonial aspects of medical anthropology.

In the Excursus I explore the construction of the category of the Third using the North American native notion of the *berdache* as an example. The Third reappears in cross-cultural research on sex-gender-sexuality-systems and is used as a term to describe everything that deviates from their Western conceptions of normal maleness/masculinity, femaleness/femininity, and heterosexuality. By analyzing the uses of the Third as an umbrella term for homosexuality, inter- or transsexuality/genderism, I show that it is always already an empty signifier that construes the supposedly a-/normal. In the case of the two-spirit movement the empty signifier of the *berdache* (as a Third) has been reclaimed, dismissed, and filled with a new meaning. I argue that Native American two-spirits found a way to disassemble their colonial history. Through a re-arrangement of a multitude of voices, they create a new possible future and thereby counter neo-colonial attempts to make tow-spirits intelligible through a Western framework. The Two-Spirit movement has challenged the colonization of their symbolic system by Western discourse. Homi Bhabha's concept of the third space and Marjorie Garber's use of the third can be sees as a disruption of the essentialist notions fostered by particular constructions of the Third in anthropology and is explored in light of the Two-Spirit movement's challenge to Western anthropology.

Chapter 6 continues to interrogate Herdt's anthropology exemplarily as reflecting the history of the discourses of evolution, psychoanalysis, and sexology. The discursive framework from which cross-cultural intersexualization originates can be traced in these disciplinary interconnections. The Sambian *kwolu-aatmwol* (hermaphrodite)—what Herdt understood as a third gender—fascinated him. He used Freud's term polymorphous perverse to explain the societal organization of the Sambia as well as their permissiveness towards a third sex_gender. Freud's term emerged within evolutionary discourses dominant at the turn of the century and thereby reflects anxieties about the permeability of racial and sexual borders at the time. In the nineteenth century, hermaphroditism was a sexual degree analogous to a racial degree. Each degree occupied an intelligible place in the natural (read: colonial, patriarchal, heteronormative) order. In the revived notion of the fifth other as an addition to the four privileged objects of knowledge by Foucault (1978) of the sexual savage in cross-cultural intersexualization, these orders are reinstalled. Cross-cultural intersexualization is deeply influenced by scientific tropes and metaphors that derive from Darwinist discourses and Freudian psychoanalysis. These discourses link the body and the mind to discourses on maturity. The civilized and sophisticated West can rest and look upon the other and their other sexuality, namely that of arrested development or DSD. Nothing is threatened; only the bodies of intersexualized people are continuingly mutilated to fit the norms of the Western sex-gender-sexuality-system.

In the conclusion I suggest, that intersexuality is at the core of the process of the construction of a dichotomously sex(ualiz)ed_gendered society. The category of intersexuality is repeatedly used to re-articulate and argue for a distinction between male and female and masculinity and femininity, as well as for a distinction between homo- and heterosexuality. The harm produced by the inauguration of this category in medical discourse and practice is atrocious and recently emerged as a human rights discourse.[3] Intersex activists speak out and resist the patronizing epistemological imperialism and mutilating violence of the medical establishment. The recent colonization of intersexualized bodies by dismissing intersex as an identity, which could and has actually claimed recognition outside of the medical establishment, was achieved by replacing intersexuality with DSD. Since the late 1990s, the emergence of intersex activism and the creation of an intersex movement critical of the Western medical establishment has made it possible to develop new and positive self-understandings that eschew scientific narratives of pathologization (e.g., Alexander, 1999; Davis, 2015; Dreger, 1999; Karkazis, 2008; Kessler, 1998).

From Myth to Medicalization

At the beginning of the twenty-first century, intersex activist Kiira Triea uses the email signature "don't quote Ovid to me" in her correspondence. In the *Metamorphoses* Ovid (43 BC–17/18 BC) told the mythology of Hermaphroditus: the son of the gods Hermes and Aphrodite. While Hermaphroditus bathed in a fountain, the nymph Salmacis fell in love and consequently wished to be united with him. The gods heard her desire and fulfilled it by joining the two together forever. Hermaphroditus from then on lived with Salmacis as a part of him. Some interpretations of this myth even state that Hermaphroditus was only 'half a man' from that point onwards (Brisson, 2002). This story is interesting in many respects. It is a highly violent narrative, which depicts the colonization, or invasion, of one body by another, of one entity by another. It also implies that two initially separate entities are now joined in a new hybrid form. The myth depicts a specific engagement with the human body and its place in the mythological order of things. In the history of intersexualization in the West, the human body has taken on different places in the socio-political sphere before entering the sphere of the medical. What Triea might mean with her email signature is that, as a contemporary intersex activist, she is tired of hearing that the problems she faces in this society have existed for several thousand years. Additionally, she may feel repulsed by the implication of a coherent narrative that assigns to her a specific, limited, and inherited, determined place in society. Moreover, a supposed mythical and ancient component provides an explanation for the atrocities to which intersex people were subjected to at the beginning of the twenty-first century. This storytelling of the supposed origin of hermaphroditism/intersexuality is therefore dismissed by Triea as a frame of reference: it cannot provide us with any valuable information on intersex since it is and remains a myth. In fact, it consolidates the notion of intersexuality as a mythical 'phenomenon.' Thea Hillman states that "while the myth of Hermaphroditus has captured the imagination for ages, it traps real human beings in the painfully small confines of story. Someone else's story" (Hillman, 2008: 2). To reveal the production of intersex(ualiz)ed bodies by showing the shifts that have taken place, is to counter a unified and coherent narrative of the subject. The phantasm, not the myth, has to be historicized, opened up, and made approachable to different interpretations.

The traceable process of medicalization is important here. Medicalization is the process through which something is made into a medical problem, to be treated, analyzed, diagnosed, and solved by so-called experts. Of course, the process becomes more powerful the more so-called experts problematize the issue and show epistemological interest. The following paragraphs explore the (self-)creation and emergence of so-called expertise in history and the creation of hermaphroditism/intersexuality as a problem. I examine historical accounts ranging from ancient Greek mythology to stories from the nineteenth century while focusing on historical accounts that are concerned with hermaphroditism/intersexuality and their explanations, be they divine or medical. These accounts are organized chronologically and focus on major works ranging from Aristotle to Hirschfeld and Freud. I am especially interested in how these works have been translated, interpreted, and made intelligible. In order to understand this I trace major shifts in the conceptualization of hermaphroditism/intersexuality and locate them in their specific time frames. The body as an object of knowledge has experienced major shifts in its conceptualization as healthy, sick, male, female, normal, abnormal; additionally, the body has been subjected to a variety of frameworks through which it has been interpreted and represented. The ability of bodies to reproduce has fascinated researchers and writers in the West[4] for centuries: their attempts to explain and represent their findings are presented to the audience of later centuries. In the following section, I depict this authoritative train of handed down accounts on the body relating to the processes of intersexualization.

Hippocrates of Kos (*a.*460–375 BC), a famous physician during antiquity, assumed that sex_gender[5] existed along a continuum from the extreme male to the extreme female with hermaphrodites located in between these two states. Of particular importance to the Hippocratic view of hermaphroditism was the idea of 'being in between the sexes.' Hermaphrodites were therefore regarded as intermediate: neither male nor female. Hippocrates considered the sex_gender of the fetus as determined by two opposites: the maternal and the paternal principles generating different seeds. The fetus would inhabit a position on a spectrum depending on its own position in the womb and the dominance of the seed. This spectrum ranged from the unambiguously female to the unambiguously male both located on different sides of the uterus. Female offspring were situated on the left side and males on the right side. In addition, he considered both to be produced by either female or male seeds, donated by both parents. Every other combination was thought to produce an intermediate sexual nature on the spectrum, either effeminate and fragile males or strong and masculine females. In the event of balanced male and female components, the fetus was located truly in the middle of the uterus and therefore hermaphroditic (see e.g., Krämer, 2007; Schleiner, 2000). The physician Claudius Galen von Pergamon (131–201 AD) revived the Hippocratic models and modified them slightly to accommodate the one-sex model which explained the sexes_genders as being caused by humors and heat. Most historians, therefore, speak hereby of the Hippocratic-Galenic model (Klöppel, 2010; Laqueur, 1990).

Alternatively, Aristotle (384–322 BC), a student of Plato, did not consider the hermaphrodite as a being of intermediate sex but rather as a being with redundant or double genitalia. Aristotle, who denied the existence of the female seed, declared that the hermaphroditic fetus was produced when the 'nurturing matter,' contributed by the mother, was more than enough for one but insufficient for two fetuses. The hermaphroditic birth appeared when either the maternal matter or the paternal seed had not been fully mastered by the other; but, if the seed mastered the matter in one part but not in the other the fetus would have both sets of genitals. For Aristotle, hermaphroditism was a condition only of genitals, "like extra toes or nipples, in that it represented an overabundance of generative material" (Dreger, 1998: 32). Therefore, from many accounts of this period it can be concluded that hermaphroditic beings were a recognized, if not fully accepted, part of ancient Greek and Roman societies. Sometimes, however, hermaphrodites were perceived as monsters and put to death or exiled. Both the Hippocratic-Galenic and the Aristotelian model survived to varying degrees during the following centuries in medical accounts. They were to differing degrees modified and adapted to the medical literature (Klöppel, 2010).

In the following centuries, *teratology*, the science of monsters, emerged. *Teratology* regarded monstrous births as omens, predictions or divine warnings (Bates, 2005). The Middle Ages witnessed a change in the perception of monstrosities: away from the consequences of carnal indulgence and bestiality to "divinations, forebodings, and examples of the wrath of God, as well as forms of glorification of God's might and power" (Dreger, 1998: 23). Conceptions of sex differences and gender hierarchy were influenced mainly by the hegemonic discourse of Christianity. Subsequently, the gendering of human beings and even of the world as a whole was produced and justified through the Bible. So-called monstrous beings were often killed on the grounds that, as Dreger reveals, "the 'monster' was a supernatural portent, a messenger of evil, a demonstration 'of bad happenings,' and as such it deserved and even required prompt annihilation" (Dreger, 1998: 33). The body was regarded as an instance of the sacred whole, a register of the cosmological order. Every being was considered to have a place in the logic of the world (Platt, 1999). The body was essentially seen as a rational one, which replicated the larger cosmology and was both sacred and universal. In the twelfth century, "the hermaphrodite seems to have represented an intriguing intellectual problem in an age determined to draw clear distinctions and boundaries—logically, politically, culturally, and socially" (Nederman & True, 1996: 515). The following Early Modern Period was characterized by the attempt to understand the body empirically, mostly through making it profane. The anatomical body, understood in structural terms, emerged in that period.

For the Early Modern Period, Levin and Solomon state that:

> the once sacred body, surrounded by cultural taboos, suddenly became a worldly machine, a matter of interiority, a profane flesh to be seen into and seen through, a presence conceived as if its mechanisms would eventually be transparent for technological knowledge.
>
> (1990: 519)

Although a shift took place, a suddenness of change as implied here is arguably misleading. The sometimes slow and highly contradictory changes need to be examined carefully as various researchers have done since the late 1990s (Epstein 1990; Jones & Stallybrass 1993; Klöppel 2010; Long 1999; Reis 2009). Moreover, social and cultural shifts started to influence the bio-medical discourse. The basic transformation of the order of things in the Early Modern Period occurred at several levels of culture and society. The social order became highly unstable because of changes in socio-economic organization, banking, and trade, and changes in the demand for capital and credit. The Reformation, the breakdown of rural community structures and other factors had a great influence on a new individualizing process. From the middle of the sixteenth century on, notably since the publication of Ambroise Paré's *On Monsters and Marvels* in 1573 (Paré 1982), a series of complex and contradictory events occurred which seemed to indicate the emergence of sex/gender as a crucial site for ordering and reordering knowledge on society and nature. The sex(ualiz)ed_gendered individual emerged in this period, and hermaphrodites surfaced in this new order of things as subjects with a natural role in society. In the Early Modern Period, medicine did not exercise hegemony over other discourses such as literature, politics, religion, and jurisprudence. Furthermore, medicine itself was hardly unified on this issue. Rather, a coexistence of alternative conceptions of the body and heterogeneity within multiple medical and juridical accounts characterized the sixteenth, seventeenth, and eighteenth centuries. However, the most basic distinction one can make is between the Hippocratics and the Aristotelians. Both schools differed in their theories of hermaphroditism as well as their above-mentioned notions of sex(ual) differences. Their contrasting accounts of hermaphroditism were transmitted into the Early Modern Period and were woven into the fabric of medieval and early modern medicine. Many accounts of hermaphroditism were written during the Early Modern Period.[6] Hermaphrodites were objects of intense speculation and interest. Cases, causes, classifications and especially status were broadly discussed. Lorraine Daston and Katherine Park, in their article "The Hermaphrodite and the Orders of Nature" suggest a unique fascination with hermaphroditism in the Early Modern Period and place the hermaphrodite "within new explanatory frameworks and linked with new fields of gender associations during this period" (1995: 419).

These various accounts can be read as the expression of a moral and social urgency concerning hermaphroditism, but they can also be placed within the tradition of ancient (later revived in the Renaissance) and medieval reflections on reproductive and sex(ual) differences. Older models still exerted great influence on the mode of investigation of hermaphroditic bodies but were increasingly undermined by early modern methodologies. sixteenth century thinking was still dominated by the Aristotelian notion of gender and reproduction, although a revival of Hippocratic medicine developed towards the end of the century. These two medical authorities, which can both be described as naturalistic, had different implications for sexuality and gender. Galen's conceptualization of medicine, which was located in the Hippocratic tradition, regarded heat as the central determining concept. Within this theoretical framework, female organs were described as a version of the penis. The female hermaphrodite was generally understood to be a woman whose 'member' enabled her to penetrate another woman's vagina. This member was explained as a vagina, which 'popped out' because of increased body heat. Sex_gender was therefore a manifestation of heat. In this tradition, sex_gender was not necessarily fixed at birth, but was unstable and could be changed during a person's life. Despite the explicit insistence of this model on male supremacy, the unisexual model served to unfix the body of two-sexed_gendered categorical restraints (Laqueur, 1990). Men were understood as having passed through a female developmental phase; this notion precluded a system of sexual dimorphism in the Early Modern Period. The revived authority of the Aristotelian notion of sex_gender in relation to redundancy or affluence is found in another dominant medical model of the Early Modern Period. The most notable difference between these two perceptions is that Hippocratics saw the cause of hermaphroditism in the entire organism whereas writers in the Aristotelian tradition saw hermaphroditism as the product of a local excess of matter and imbalance of male and female principles. For them male and female were not points on a spectrum but bipolarities which could not be mediated. In conclusion, one could say that the Hippocratic model of a sex_gender continuum posed a potential challenge to the male-female bipolar dichotomy, and therefore to the entire social order, whereas the Aristotelian model never questioned this order by viewing the existence of ambiguous bodies as superficial and leaving the dichotomous order intact. Both models adapted elements from each other (Laqueur, 1990) and have been historicized as interchangeably dominating each other (see also Krämer, 2007).

The mixture of these two different medical models can be observed in Ambroise Paré's *On Monsters and Marvels* (Paré, [1573] 1982). He operated from both approaches to the human body to explain the case of a woman who suddenly changed into a man. Transgressing the bounds of gender, with resulting inappropriate behavior, could cause a change of sex. This is observable in different recorded cases, for example that of Marie who turned into Germain. This occurred during Paré's lifetime in France, in Vitry-le-Francois between *c.*1510–1590. Marie had lived and dressed like a girl until the age of 15, then in the heat of puberty, the girl who had "no mark of masculinity" was chasing pigs and jumped across a ditch and "at that very moment the genitalia and the male rod came to be developed in him, having ruptured the ligaments by which previously they had been held enclosed and locked in" (32). After attending physicians and surgeons, who "found that she was a man, and no longer a girl," Marie became Germain: she put on men's clothes and from then on lived as a man. For Paré

> the reason why women can degenerate into men is because women have as much hidden within the body as men have exposed outside; leaving aside, only, that women don't have so much heat, nor the ability to push out what by the coldness of their temperament is held as if bound to the interior.
>
> (32)

The term *degenerate* here does not imply that men were less perfect than women, yet bodily boundaries are described as fluid. However, at this point, Paré refers to the Hippocratic (Galenic) notion that heat causes the interior female genitals to externalize into male genitals. Yet, in the following explanation he clearly uses the Aristotelian understanding of the female to be an imperfect or defective male. Paré states that:

> now since such a metamorphosis takes place in Nature for the alleged reasons and examples, we therefore never find in any true story that any man ever became a woman, because Nature tends always toward what is most perfect and not, on the contrary, to performing [in] such a way [that] what is perfect should become imperfect.
>
> (33)

In Marie/Germaine's case, jumping allowed the internal genitals to drop out. In other cases puberty or active sex, or anything inappropriate to women was reported to have increased the heat in the cold bodies of women (Parker 1993). This increase in heat was said to have caused them to become men. All women were thus potential men. However, the threat derived from this interpretation of the body did not yet gain its full strength; it would gain it later on. Such possibility of women crossing the sex_gender line maintains its threat through the centuries, although the responses vary.[7] A bit more heat or acting the part of the other sex_gender could suddenly bestow a penis, which entitled its bearer to the mark of the phallus, to be designated a man. According to Laqueur, these changes:

> in corporeal structures, or the discovery that things were not as they seemed at first, could push a body easily from one juridical category (female) to another (male). These categories were based on gender distinctions—active/passive, hot/cold, formed/unformed, forming/formable—of which an external or an internal penis was only the diagnostic sign. Maleness and femaleness did not reside in anything particular. Thus for the hermaphrodites the question was not 'what sex they are really', but to which gender the architecture of their bodies most readily lent itself.
>
> (1990: 135)

In the Early Modern Period intense attention was therefore devoted to hermaphroditism; it became associated with sexual, moral and theological issues clearly linked to the subjects of sodomy, transvestism, and sexual transformation. These subjects threatened the orders of society in the Early Modern Period because they blurred the social distinction of hierarchical gender roles, which were necessary to be upheld in all these ruptures of the order of things. Paré formulates this threat:

> For some of them have abused their situation, with the result that, through mutual and reciprocal use, they take their pleasure first with one set of organs and then with the other: first with those of a man, then with those of a woman, because they have the *nature* of man and of woman suitable to an act.
>
> ([1573] 1982: 27)

After his remarks about hermaphroditism, Paré immediately moves to a discussion of sex between women. This association of sex and sodomy was characteristic for how the term hermaphroditism became clearly associated with sex_gender related subjects such as male and female sodomy[8] and transvestism. It seems as if the term hermaphroditism became associated with practices that appeared to blur or erase the lines between the sexes_genders; it became emblematic of all kinds of (sexual) ambiguity. Nevertheless, during the sixteenth century it was still assumed that nature would show itself in the mature hermaphrodite through attraction to the 'opposite' sex. Therefore, it was left to the person in question to decide in which role they wanted to live: no medical or other authoritative consultation was required. Sex_gender was not predetermined through bodily boundaries based on medical investigation. Sex_gender was not a lifetime fixity but a temporary examination of appropriate sex_gender behavior. According to Laqueur, the shift from the *one-sex model* to the *two-sex model* took place during this time. He notes that during much of the 17th century, to be a man or a woman was to hold a social rank, to assume a cultural role, and not to *be* organically one or the other of two sexes_genders. Sex was still a sociological, not an ontological, category.

(1990: 142)

Therefore, a hermaphrodite could be regarded as having two sexes, between which s_he[9] could make a social and juridical choice; there was no true, deep essential sex_gender that differentiated the cultural woman from man.

With the seventeenth century, however, came an increase in the reliance on outside testimony to determine the hermaphrodite's predominant sex_gender. The threat that derived from this cross-mixture of sex_gender boundaries is obvious in the case of Marie/Marin that appeared in 1601 in France. The woman Marie publicly declared at the age of 21 that she was in fact a man, changed her name to Marin, put on men's clothing and announced his intention to marry another woman. The subsequent prosecution and sentencing to death, based on the accusation of sodomy, was then relaxed after examination by a physician who argued that the (genital) member, which was initially defined as an enlarged clitoris, was rather a small penis. The death penalty was suspended on the condition that Marie/Marin wear women's clothing and remain celibate until the age of 25. This case occurs exactly at the peak of the shift from the system of distinguishing sex_gender by social factors to the system of scientific authority. According to Katherine Long, medical authors Caspar Bauhin and Jaques Duval (both publications dated 1614) insisted on medical procedures and invasive examinations to assign the proper sex related to reproductive functions and therefore the specific role in society (Long 1999). Thus, the seventeenth century brought a shift in the representation of sex_gender accompanied by a change in the modes of scientific discourse.

Regarding the accumulation of medical accounts on hermaphroditism during the sixteenth century, it can be seen that the fear of sexual fraud surrounded all forms of sexual ambiguity; therefore, these forms needed to be prohibited by investigation, definition, and classification. Daston and Park suggest that "it was the fear of sexual fraud and malfeasance, surrounding all forms of sexual ambiguity that disqualified the hermaphrodite's own testimony and demanded that of doctors, surgeons and midwifes instead" (1995: 425). The growing anxiety over so-called sexual ambiguity necessitated a different system of classification and categorization to that which featured in ancient and medieval times. Numerous explanatory systems were drawn together to find new ways of defining individual status, hierarchical relationships, constraints, and responsibilities, as well as to install a new social order founded on the production of scientific knowledge. Many historians point to this as a scientific revolution during the seventeenth century (e.g., Cohen 1994; Shapin 1998). The Enlightenment coincided with and helped to produce immense shifts in Western society. With the arrival of the Enlightenment, the power of religion and religious authority began to decline with regard to sex_gender matters. The power of the Church was challenged by new elites whose authority was drawn from the increasing prestige of the natural sciences. Historian Robert Nye gives three main characterizations of this shift:

> First, the family unit continued to be recognised as the primordal sexual space, with procreation and child rearing as its principal task. Second, the divine ordinance that made all humankind into the sons or daughters of Adam and Eve was firmly sustained by the findings of modern natural science on the differences between the sexes. And third, the newly emergent liberal and democratic order readily embraced this new knowledge and legitimised it in law, much as medieval and early modern politics had used dogma to deepen and extend their power.
>
> (1999: 67)

The heroic age of the scientist was dawning; it claimed that the power of human reason was sufficient to identify laws of science that would lead to human progress. From 1700 onwards, sexuality and therefore also sex_gender became tied to natural biological processes: they were now seen as determining the person's identity. The Enlightenment inaugurated the science of the two sexes_genders. This shift can be regarded as the development of a scientific gaze, which would become *the* authority in sexual matters from that point onwards. Historian Fabian Krämer describes this development as the new privilege of the scientific expert to diagnose and to give a name, just as the early colonists did to their 'discoveries' (Krämer, 2007: 52). Intersexualization is based on these processes in scientific expert knowledge production and had its origin in the scientific revolution.

By 1750 the increased medicalization was clearly visible in George Arnauld's *A Dissertation on Hermaphrodites* in which the surgeon categorizes hermaphrodites. He divides people into perfect, imperfect, predominantly male, and predominantly female (Donoghue, 1993: 51). Arnauld focused on genitals, and explained that a swollen clitoris can appear in many women, but only in the female hermaphrodite does it become erect and frees itself from the labia. In many cases the hermaphrodite was referred to as a woman suffering from macroclitorideus (enlarged clitoris); thus, the term implicated women who were discovered to disobey their appropriate place of the sexed_gendered society by seeking access to another status in society.[10] Furthermore, this new definition implies what a normal, heterosexual woman should look like. Clear anatomical borders became installed for the first time, whereas the figure of the hermaphrodite served as the abnormality that helped define the norm. From this point onwards, hermaphroditism was defined through 'deformed' anatomy and clearly connected to the transgression of sex_gender boundaries. Daston and Park see:

> rather a conjunction of changing medical ideas with a general climate of acute male anxiety about the very issues brought to prominence by the new interpretation of the hermaphrodite as a being of intermediate sex—especially the

issues of sodomy and other sexual crimes, and the proper relationship and boundaries between men and women.

(1995: 430)

From this moment on it was possible to explain 'anomalous beings' in terms of variations of normal and abnormal development. The domestication of the monster was achieved by the newly emerging natural sciences, whose power and authority over the matters of life increased during this time. This new ordering system of bodies colonized the 'extraordinary' body by defining, categorizing, and classifying it in relation to the natural order of things. By transferring hermaphroditism and other monstrosities from mythology into the category of pathology, the medical discourse and medical practice became the source of judgment. Monsters were now understood in terms of an analogical relationship with respect to the norm not anymore as beings related to another order. Now it was medicine, which provided reasons for tolerance or intolerance against people classified as abnormal. Hermaphrodites could serve as an anomaly reassuring the norm. Medical professionals started assigning the *true sex* of a person. Medicalization therefore teamed up with the conqueror science advancing through human bodies. Now that the natural phenomenon was named as such, it was necessary, and possible, to control and discipline it.

During the following decades of the nineteenth century medical professionals became more prestigious and more aligned with science. More people were seeing doctors and bio-medicine was on the rise, including gynecology. Increasing numbers of people had access to medical care and became subject to genital examinations. Consequently, an increasing number of hermaphrodites were discovered. The opportunities to document these cases were increasing through the growing number of medical publications at that time. The consequence was that medical practitioners became aware of the phenomenon of hermaphroditism which turned out not to be as rare as previously thought. "The years from around 1860 to 1870" as Foucault suggests,

was precisely one of those periods when investigations of sexual identity were carried out with the most intensity, in an attempt not only to establish the true sex of hermaphrodites but also to identify, classify, and characterize the different perversions.

(1980: xii)

Sex_gender therefore also became intrinsically connected to sexuality and deviance from heterosexual norms.

Charles Darwin, amongst others, laid the foundation for this shift. In the Darwinian model, sexual behavior focused on reproduction and the natural selection of males and females according to their role in both reproduction and resource competition (1859). This evolutionary biological background influenced the notion that human beings are solely organized around reproduction. The core of Darwin's thought was that the higher the organism on the ladder of life, the more exquisitely differentiated the male and the female of the species. Working within the Darwinian tradition of sexual dimorphism, late modern sexology developed the concepts of male and female as innate structures in all forms of life, including human beings. Heterosexuality was therefore teleologically necessary and regarded as the highest form of (sexual) evolution.

Before and during the nineteenth century crucial reorganizations took place in Europe. Secularization, industrialization, and warfare demanded new structures to guarantee social stability. The nuclear family emerged as a key unit of the Western nation states; it was within this nuclear core of society that the future workforce would be produced. Within this reproductive framework, same-sex_ gender desires and practices were a problem to be dealt with as they were regarded as aberrations from the procreative norm. The family was viewed as the place of reproduction, and every disturbance of this institution required revelation. Sexuality as the palladium of reproduction came under observation. The late nineteenth century was a time of anxiety in Europe. Politicians were discussing the declining birth-rate and the increasing numbers of women who never married. The first Women's Movement and Suffragettes expanded political activism and brought up the Woman Question.[11] But what was one to do if a woman was hard to define? The recently named 'homosexual' was being widely discussed and physicians "were already feeling rather worried about the instability of political-sexual identities" (Dreger, 1999: 6). Edward Carpenter's *The Intermediate Sex* (1896) depicts these social changes at the turn of the century. He states that recently, "with the arrival of the New Woman, many things in the relation of men and women to each other have altered, or at any rate become clearer" (1921 [1896]: 16). Carpenter adds that "the growing sense of equality in habits and customs—university studies, art music, politics, the bicycle, etc.—all these things have brought about a *rapprochement* between the sexes" (16). Carpenter argued for a different interpretation of the circumstances than most of his contemporaries. He actually argued for equality between men and women and seems to have been more or less immune to the Darwinian notion of exclusive sexual dimorphism. He stated:

> It is beginning to be recognised that the sexes do not or should not normally form two groups hopelessly isolated in habit and feeling from each other, but that they rather represent the two poles of one group—which is the human race; so that while certainly the extreme specimen at either pole are vastly divergent, there are great numbers in the middle region who (though differing corporeally as men and women) are by emotion and temperament very near to each other.
>
> (Carpenter, 1921 [1896]: 17)

Carpenter's courage to handle these new social challenges, however, was very rare. Sufficient evidence exists that anxiety about sex_gender roles made many physicians sensitive, especially to sex(ualiz)ed_gendered identities and therefore also to their patients' anatomy. Feminists and homosexuals who challenged boundaries at that time caused bio-medical experts to search for tighter definitions of norms. The categories of female and male had to be defined clearly in non-overlapping ways to safeguard the two-sexed_gendered, heterosexual social system. Therefore, more bodies were discovered in their nonconformity and "fell into the 'doubtful' range" (Dreger, 1998: 26). People who had no obvious and permanent *true sex* threatened natural sex_gender borders. Foucault describes the following:

> Biological theories of sexuality, juridical conceptions of the individual, forms of administrative control in modern nations, led little by little to rejecting the idea of a mixture or the two sexes in a single body, and consequently to limiting the free choice of the indeterminate individual. Henceforth, everybody was to have one and only one sex. Everybody was to have his or her primary, profound, determined and determining sexual identity; as for the elements of the other sex that might appear; they could only be accidental, superficial, or even quite simply illusory.
>
> (1980b: viii)

Physicians were becoming unhappy with the idea that so many *true* hermaphrodites existed. Consequently, the bio-medical professionals gradually decided to tighten the classification of hermaphroditism. A reconfiguration of the category of *true* hermaphroditism became necessary in order to reduce the possibility of its existence in humans. Each case of doubtful sex had to be resolved in a diagnosis of *true* female and *true* male, which allowed the reduction of *true* hermaphroditism. Redefinition of old cases of *true* hermaphroditism as *pseudo* hermaphrodites became fashionable amongst physicians.[12] *True* hermaphrodites were literally extinguished not only from history but also in their physical appearance and ability. The physician Jonathan Hutchinson declared in 1896 that

> whenever this [true hermaphroditism] is the case, the organs are never developed in perfection. The testes do not secret semen, or the ovaries do not attain their functional activity, and this being who is in a sense bisexed is in another sense unsexed, and never attains the full development of either.
>
> (cited in Dreger 1995: 363)

This perspective was founded on the differentiation of men and women by their reproductive capabilities, originating in Darwinist theories. If people had both testes and ovaries they were *true* hermaphrodites. The gonadal definition meant that every body could officially be exclusively one sex, which was a contribution to the strict separation between males and females and hence to the enforcement of the biologically defined *one-body-one-sex* rule (Laqueur 1990).

More and more physicians became interested in the 'phenomenon' of hermaphroditism. The most notable figures were the Polish Franz von Neugebauer and the German Theodor Albrecht Edwin Klebs. By 1908 Franz von Neugebauer

(1856–1914), Director of the Gynecological Section of the Evangelical Hospital of Warsaw and a founding member of the British Gynecological Society, had collected and analyzed more than 930 accounts of hermaphroditism. His collection was published in 1906 as *Hermaphroditismus beim Menschen* (*Hermaphroditism in Man*; [my translation]) and included a bibliography on hermaphroditism. To his contemporaries, the most striking outcome of his research was the discovery of 68 marriages between persons of the same medical sex_gender. In light of the historical developments of the time this must have triggered serious concerns for the figure of the hermaphrodite: the possibility of crossing the increasingly policed sex_gender boundaries. The diaries of Alexina *Herculine Barbin* (1838–1868), edited by Foucault (1980b), exemplify the desire of the so-called medical experts to define one true sex in every person.[13]

The German pathologist Theodor Albrecht Edwin Klebs (1834–1913) created a classification system published in 1876 in the *Handbuch der Pathologischen Anatomie* (*Handbook of Pathological Anatomy*; my translation), which served to drastically decrease the number of people who could be defined as hermaphrodites. In Klebs' system true hermaphrodites had to have at least one ovary as well as at least one testicle. Moreover, he divided them further into "true bilateral hermaphroditism" (with one ovary and one testicle on each side), "true unilateral hermaphroditism" (on one side an ovary or a testicle and on the other an ovary and a testicle) and "true lateral hermaphroditism" (a testicle on one side and an ovary on the other). Finally, "false hermaphroditism," the so-called pseudo-hermaphroditism, was defined as "doubling of the external genital apparatus with a single kind of sexual gland" (Klebs 1876, quoted in Dreger 1998: 145). This "false" hermaphroditism was further divided into two separate categories, the "masculine pseudo-hermaphrodite" with testicles and female genitals, and the "feminine pseudo-hermaphrodite" with ovaries and masculine genitals. Klebs thus reinforced the popular conception that there were two and only two sexes_genders, with a rare and unusual exception in the case of true hermaphroditism. Alice Dreger quotes Halliday Croom, who emphasized the "importance of making early discovery of such cases, in order to save the miserable consequences, unhappiness, and divorce suits, even suicides, which may follow if they are not recognised and are allowed to proceed in error" (1995: 341). The question remains whether this was meant to be for the sake of 'the patient' and the partner. In general, as Geertje Mak puts it,

> instead of offering the hermaphrodite the right to choose his or her own sex, they [the physicians] started to turn sex-gender consciousness into an object of medical investigation, into a measurable identity whose importance in relation to the final decision only they could define
>
> (2005: 87)

Along with the continuous redefinition of the essentials that made a true hermaphrodite, the essentials that made true males and true females were also changed. The emphasis on reproduction during the end of the nineteenth century brought immense attention to issues of sex_gender and sexuality.[14] The 'anatomical hermaphrodite' challenged the traditional image of two distinct sexes_genders and the newly construed invert—the hermaphrodite in the soul—threatened the social heteronormative order of the nuclear family.

Following from the deep concern with sex(uality) during this time and a general professionalization of the sciences (such as medicine, biology, psychology, anthropology, and so on), sexology became a scientific discipline. The pioneers of sexology such as Magnus Hirschfeld, Havelock Ellis, and Richard von Krafft-Ebing, were located in Germany or in the United Kingdom. They were highly invested in sexology in general[15] and homosexuality[16] in particular. Interestingly, most of them tried to foster public acceptance of homosexuality. The way they chose to reach this goal was by naturalizing homosexual desire. The category of sex became the foundation for sex_gender dichotomization and the stabilization of heteronormativity. The 'psycho-sexual hermaphrodite' became a category of a mental state and was used to construe an 'invert.' What I call 'diagnosticism,' the compulsion to diagnose, features in the invention of the category of the homosexual as much as it features in the processes of intersexualization (Eckert, 2009).

When Karl Heinrich Ulrichs published his first accounts on *Uranism* (his neologism for the invert) in 1868, he spoke about a *third sex*, which was hermaphroditic in the soul, not in the body.[17] Growing evidence of homosexuals' anatomical normality, however, challenged the hermaphroditic model imposed by advocates of the third sex_gender, and the emphasis placed on the body shifted to the mind. In the following decades, the terms hermaphroditism and homosexuality became intermingled. The notion of the anatomically deviant transformed into the psychologically deviant. Homosexual inverts came to be seen as masculine woman or effeminate man, according to their sexual orientation. In the notion of Darwinism, hermaphrodites *and* homosexuals were considered to be unfinished specimens of stunted evolutionary growth.

The rise of a new era of explanations of the world order had begun. Social phenomena and the development of civilization became located in a natural order of things. At the end of the nineteenth century, Darwinism was accepted and the idea of evolution with man at the top was established. Every being was considered to have a place in the evolutionary process of creation. Progress was signified by the greatest degree of sex(ual) difference, as well as procreative heterosexuality. The notion of sexual dimorphism as the pride of creation makes these ideas so crucial for the perception of hermaphroditism. In the tradition of Darwinian theories on the evolution of organisms through natural selection, the differentiation between the sexes became a sign of an evolutionary progress towards civilization.

Richard von Krafft-Ebing presented congenital sexual inversion in his famous *Psychopathia Sexualis* (1903), the grand encyclopedia of sexual perversities according to Rosario (1997: 15) in four gradations. His classification system depicted the mixture of two concepts: the anatomical and the psychological. He distinguished between: (1) psychical hermaphroditism, where subjects are mainly homosexual but traces of heterosexuality remain; (2) homosexuality, where there is an inclination towards members of the same sex only; (3) viraginity (in women) and effemination (in men) which means that the invert's psychical character corresponds completely with the invert's sexual instinct; and (4) hermaphroditism or pseudo-hermaphroditism, where men's bodies become feminized and women's bodies masculinized and the subject's physical form begins to correspond to the inverted sexual instinct (Krafft-Ebing 1903). The influence of this kind of understanding of hermaphroditism is also apparent in Magnus Hirschfeld's theories.[18] He published the *Jahrbuch für sexuelle Zwischenstufen* (*Annual for Sexual Intermediaries*), which advocated the theory that all subjects were, to varying degrees, bisexual and/or transgendered and that sex was impossible to be categorized in any oppositional sense (1899–1923).[19]

However, sexology as a newly developed science was concerned with the belief that certain socially disadvantaged groups of people were intellectually inferior by nature. Thus, the bodies of, for example, the poor, criminals, non-white people and women were assumed to be primitive and diseased. After the invention of the category of the homosexual, the body that expressed deviant sexual desires joined this group of suspected degenerates. The invention of categories of distinctive sexual types (as Foucault has shown for the end of the nineteenth century) was continued in this tradition. Richard von Krafft-Ebing, in particular, brought the notion to the fore that same-sex desires were not just behaviors but inherent in the individual. He believed in the hereditary basis of inversion but he also thought it should and could be cured. His actual intention was to strengthen the 'natural' status of homosexual men and women. Havelock Ellis also proclaimed 'the organic nature of inversion' yet he assumed it to be curable. This was in contrast to Krafft-Ebing, the first to treat "homosexuality as neither a disease nor a crime" (Grosskurth 1980: 185). Krafft-Ebing made a distinction between perversity and perversion which were defined as vice and disease respectively (1903). Therefore, sexologists tried to fight the pathologization and criminalization of homosexual desire by naturalizing it. Sexologists described homosexuality as a product of hereditary degeneration and presented it as a harmless variation in human sexual behavior. The effects of this scheme extended into the following decades; biological determinism came to be a central issue in homosexualization and intersexualization. The quest to define homosexuality as an inherited bodily attribute peaked with the investigations of the *Committee for the Study of Sex Variants* that was conducted from 1935 to 1941 in New York. The participants of the study were volunteers with a history of homosexual relationships and were sent through various psychological and physiological examination processes. Jennifer Terry describes this case concerning the construction of the lesbian body. The researchers believed that genitals revealed masturbation, frigidity, promiscuity or lesbianism. They investigated and drew graphic sketches of female genitals to prove that the deviance shows itself in the bodily condition. Terry concludes that genitals became "indices of moral character" (1995: 143). This study depicted the 'lesbian clitoris' as a clear threat to the actual 'natural phallus' in men as similar in size and shape by using the term hypertrophy which would remain to describe hermaphroditism/intersexuality throughout the following decades. This specific case reveals how the hetero-sexual matrix wove itself into the perception of somatic features. It was founded on the following questions: What is acceptable concerning the make-up of genitals? Where is the border that can be drawn between the shape of genitals and the supposedly inherent deviance? How does 'abnormality' in the mind show itself in the constitution of the body? This study reveals the changing status of the body as scientific proof of homosexuality. For the following decades the scientific quest to fix identity on bodily features would continue. In intersexuali-zation it becomes especially clear that genital features acquired the status of a signifier for the threat to the 'normal' heterosexual world.

According to Gender Studies scholar Steven Angelides, the period of early sexology was marked by "the invention of the category of 'sexuality' in general, and the opposition of hetero/homosexuality in particular" (2001: 17). Sexology with the invention of sexuality as a dominant feature of human nature completed the developments, which Foucault has described regarding the nineteenth century, in which he saw the notion of sex emerging as the secret:

> The notion of "sex" made it possible to group together, in an artificial unity, anatomical elements, biological functions, conducts, sensations, and pleasure, and it enables it to make use of this fictitious unity as a causal principle, and omnipresent meaning, a secret to be discovered everywhere: sex was thus able to function as a unique signifier and as a universal signified.
>
> (1978: 154)

It was not just the homosexual threat that needed to be reduced but sexuality suddenly became a field of study: thus, the academic discipline responsible for meeting this threat. However, by contributing to the pathologization of the her-maphrodite/intersexual, it unquestionably represented a continuous link between the construction of norm and anomaly. The deviant body of the homosexual is of high importance for the category of the hermaphrodite in this period. Hermaph-roditism served as a tool to describe homosexual desire. Common terms were hermaphroditism of the mind or psychic hermaphroditism, although there was growing evidence for normal bodies of homosexualized people. However, the hermaphrodite became the signifier for two different kinds of deviance: first the disturbance of clear sex_gender distinction; and second, the trouble this phenom-enon caused to the desired clear split between hetero- and homosexuality. Just as heightened attention was paid to the hermaphrodite, the term of bisexuality in relation to sexual desire or orientation emerged. This is *the* emblematic develop-ment in sexology talk connected to the moment in which homo- and heterosexu-ality were invented. As Angelides puts it: "whether explicitly defined or not at the moment of homo- and heterosexuality's scientific intervention, the notion of a dual sexuality, let us call it bisexuality, is without doubt a logical or axiomatic component of such a dualistic structure" (2001: 15). This dualistic structure was two-fold, a double bind system composed of the compulsory conceptualization of one as "both a man *and* either a heterosexual or a homosexual, a woman *and* either a heterosexual or a lesbian" (24). The concepts of the hermaphrodite/intersexualized and that of homo-/bisexuality mark the epistemological alliance of sex_gender and sexuality.

Sigmund Freud's psychoanalytical theories of sexuality influenced the scientific community and shifted the focus from the body to the psyche (although the notion of somatic qualities of the homosexual was still held up). Freud's theories on innate bisexuality triggered a specific notion of hermaphroditism as an originary stage in human development. Freud established the connection between the psyche and the body in relation to civilization in general and to maturity on the individual level. Freud's theories relied heavily, just as most of his contemporaries' theories, on a Darwinian perspective of life and social organization. In Freud's psychoanalytical accounts, however, the development of the psyche came to be linked to the body (though arguably more or less intensely and with different emphasis throughout his works). Freud's earlier writings were significantly more open to the fluidity and openness of psycho-sexual development and only later became focused on a rigid heteronormative organization of embodiment, subjectivity, and desire. The researchers whose work is interrogated in this book focus exclusively on the later writings and thereby re-install static, deterministic, and essentialist notions. The heteronormative and dichotomous sex-gender-sexuality-system, which they reaffirm time and time again is based on a specific reading of Freudian psychoanalysis and erases another.

Richard Goldschmidt (1878–1958) coined the term 'intersexuality' in his genetic sex_gender determination theories. He stated that 'bipolar' sex_gender difference exists but would only manifest itself on the basis of a "bisexual potentiality" (my translation, German: *bisexuelle Potenz*) and only as a "quantitative" and therefore a "relative" differentiation (Goldschmidt, 1916, 1938; see also Klöppel, 2010). The highly influential and widely read *The Evolution of Sex* by Patrick Geddes and Arthur Thomson from 1889 stated that "hermaphroditism is primitive; the unisexual state is a subsequent differentiation. The present cases of normal hermaphroditism imply either persistence or reversion" (80). The notion of natural selection made it possible to view hermaphroditic bodies as anomalous 'throwbacks.'[20] Therefore, the hermaphrodite came to be seen as atavistic, as unfinished in its development yet as having a place in the natural order of things albeit in need of normalization. This phenomenon will be of central interest in my analysis of the following decades.

This brief historicization of the early processes of intersexualization has shown how the foundation of heteronormativity became justified by science. With the rise and spreading acceptance of naturalization/normalization, what Foucault described as bio-power, was now institutionalized by the newly emerging disciplines, which defined every-body into the spheres of the normal and the abnormal. Foucault, in *Discipline and Punish* characterizes disciplinarity as an atomizing force:

Instead of bending all its subjects into a single uniform mass, it separates, analyzes, differentiates, carries its procedure of decomposition to the point of necessary and sufficient single units. It "trains" the moving, confused, useless multitudes of bodies and forces into a multiplicity of individual elements—small, separate, cells, organic autonomies, genetic identities and continuities, combinatory segments. Discipline "makes" individuals; it is the specific technique of a power that regards individuals both as objects and as instrument of its exercise.

(1977: 170)

The knowledge of embryology, biochemistry, endocrinology, psychology, sexology, and surgery enabled physicians to control the matters of sex_gender and sexuality in regard to social organization. The institutional authority over deviance was now clearly located inside the medical establishment. As the increasing discourse of bio-power gained influence, the hermaphrodite became fully naturalized.[21] Unfortunately, the aim of sexologists to make homosexuality accepted only reinforced the dichotomy between normal and abnormal, between heterosexuality and homosexuality, between natural and unnatural. The hermaphrodite played an important role, understood either as a bodily condition or as a mental condition (newly defined and modified according to the newly appearing emergencies of social and cultural changes). The body of the hermaphrodite/intersexualized was the playground on which scientists fought their battle. The perception and importance of hermaphroditism/intersexuality was, as a result of social circumstances, shifted to the field of sexual inquiry through a purely social or legal interest. A new dichotomy was created at the cost of the hermaphrodite. Since then, there is not only the dichotomy between male and female, but also a new one: the dichotomy between homosexual and heterosexual. In the following decades, the hermaphrodite suffered two fights, one against the blurring of the sexes_genders and one against the blurring of 'straight' desire and sexual practices represented by the terms of bisexuality and hermaphroditism merging. The term intersexuality was coined and freed the hermaphrodite from any mythological or religious background for its scientific investigators. Medicalization was completed—intersexualization on the rise.

Notes

1. The * stands for the options which are available after the inter, such as—sexual, -sexuality, or gender. These variations have recently emereged mainly due to activist reclaiming and defying of terms that the medical establishment uses.
2. This is done in line with the author s_he who published an article called "Performing the Gap—Queere Gestalten und geschlechtliche Aneignung," the German magazine *arranca*, 28 (http://arranca.org/ausgabe/28/performing-the-gap).
3. To my knowledge, the exhibition *1–0–1 [one 'o one] intersex* at NGBK in Berlin Germany in 2005 was the first to title: *"Das Zwei-Geschlechter-System als Menschenrechtsverletzung"* (The binary sex_gender system as a human right violation [my translation], exhibition catalog: NGBK 2005). In 2005/6 *Cardozo. Journal of Law and Gender* published a special issue on intersexuality and human rights discourses (e.g., Benson, 2005; Bird, 2005; Gruber, 2005; Holmes, 2005).

4. I do not want to indicate that it is only the West in which researchers pursued this interest. However, in this book my focus is on the epistemological space of the West and does not claim to make any statement about knowledge production outside this particular epistemological space.

5. Because historically there has not been a distinction between sex and gender I use the _ underscore between sex and gender to denote that both terms are inextricably inter-linked in current understandings of the issue.

6. To mention a few: 1600, *De Hermaproditorum monstrorumque partum*, Caspar Bauhin; 1612, *Discourse on Hermaphrodites*, Jean Riolan; 1614, *Treatise on Hermaphrodites*, Jaques Duval; 1614, *On the Nature of Births of Hermaphrodites and Monsters*, Caspar Bauhin; 1642, *Monstrorum Historia*, Ulisse Aldrovani; 1653, *Questionum medico-legalium*, Paolo Zacchia; 1671, *The Midwives Book*, Jane Sharp; 1692, *Discursus juridico-philolocus de hermaphroditis*, Jacob Möller (see Klöppel, 2010, Krämer, 2007).

7. As will become clear, the threat of women crossing the gender line will later on also be banned by the means of surgery. Most intersex* surgeries are performed to prevent a person living as a women having a penis.

8. 'Female sodomy' is not a synonym for the term 'lesbianism.' As Foucault has shown, the category of the homosexual only emerged at the end of the nineteenth century as a category of identity. Before this time, 'female sodomy' only described female same-sex_gender acts. The term tribadism would also be used, albeit not as an identity category as which 'lesbian' would be used in the twentieth century.

9. I use the terms s_he, him_her, and his_her to designate a space by the _ that includes people who do not feel that the pronouns or the pronominal adjectives of he/she, his/her, him/her denote their identities. This is done in line with the author s_he who published an article called "Performing the Gap—Queere Gestalten und geschlechtliche Aneignung," the German magazine *arranca*, 28 (http://arranca.org/ausgabe/28/performing-the-gap).

10. As the case of Marie/Marin has shown, Marie's 'enlarged clitoris' had to be redefined as a 'small phallus' in order for her to be allowed to live.

11. The Woman Question is here used as a general term for the political challenges of feminist movements at the end of the nineteenth century. Feminist movements differed immensely in their specific goals; however, all of them challenged the hierarchical order, discrimination, oppression, and exclusion.

12. For example the professionals G. F. Blacker and T. W. P. Lawrence (see Dreger, 1995: 34–37).

13. Alexina Herculine Barbin was raised as a girl and became a female teacher in her_his early twenties. Her medical discovery as truly male actually only happened after her_ his suicide, but medical attention was devoted earlier. This medical attention was caused by her search for support concerning her_his desire for her virginal girlfriend Sara. The seduction of Sara only became possible through "medical misdiagnosis" of Barbin's sex/gender. Contemporaries assumed that it was the inappropriate social role that enabled Barbin to get access to female spaces and it was Barbin's testes that made her_him desire Sara, a woman. Hence it was not the size of an organ but the use of it that was the problem. As Barbin insisted on behaving as the possessor of the phallus s_he could only be accepted through a reassignment as an appropriate possessor of such power: a man. With the growing acknowledgment of the definition of true hermaphroditism, Barbin came to be "legally" redefined as a true male and seen to be just a hypospade. (Hypospadias is defined as a deformity of the penis.) See also Mak, 1997, 2005.

14. Especially in Germany the protection of the 'Volkskoerper' (the nation body) was discussed as being threatened by non-reproductive sex and it came to concern also economical productivity in general. Eugenic discourses were fed by sexological knowledge production.

15. Sexology developed expertise on subjects such as venereal disease, eugenics, sexual psychopathology, prostitution.
16. The homosexual at that time was only discussed as appearing in men, lesbianism was hardly discussed, although Havelock Ellis presumed that female homosexuality was more common than in males but it was not regarded as a 'social evil' as it was in men.
17. Hirschfeld later modified Ulrichs' theories on the intermediaries (in 'sexual orientation') for his categorization of hermaphroditism in the *Jahrbuch für sexuelle Zwischenstufen*.
18. Hirschfeld founded the "Scientific Humanitarian Committee" (Wissenschaftlich-Humanitäre Komitee [WHK]) in 1897 and used it as a medium to bring his struggle for homosexual emancipation to the fore.
19. The rise of the homophile movement advanced the circulation of modern sexologist theories. The main attempt was to seek tolerance for homosexuality, for instance by linking it to hermaphroditism (as a pathological condition) or inventing the category of the third gender (in terms of sexual object choice). Early sexologists viewed homosexuality as an innate constitutional condition, but not all of them considered it pathological. Jennifer Terry and Jaquline Urla state that "their most pronounced scientific legacy was the idea that the homosexual was an inherently different type of person, endowed with somatic and characteriological features that distinguished this creature from normal people" (Terry & Urla, 1995: 137).
20. Geddes and Thomson discus Darwin's theory of sexual selection in full length in the first chapter of their book (1889: 3–31). I will elaborate on the influence of social Darwinism on sexology and psychoanalysis in Chapter 6.
21. Naturalization means that natural science is the medium that can explain the natural law in which the hermaphrodite is settled. Naturalization means the process in which an uncertainty is rendered unproblematic or 'natural' or self-evident. In the context of intersexualization naturalization is a basic feature, which interlocks with normalization in the sense that that which is seen as 'natural.' Sexual dimorphism is that which is regarded as natural, however, it is constructed by means of surgical intervention to reaffirm that hermaphroditism/intersexuality is an 'experiment of nature' and therefore needs to be normalized.

Bibliography

Adams, W. L. (2009, August 21). Could this woman's world champ be a man? *Time*. Retrieved March 14, 2016, from http://content.time.com/time/world/article/0,8599, 1917767,00.html.

Alexander, T. (1999). Silence = Death. In A. D. Dreger (Ed.), *Intersex in the age of* ethics (pp. 103–109). Hagerstown, Maryland: University Publishing Group.

Angelides, S. (2001). *A history of bisexuality*. Chicago, London: The University of Chicago Press.

Banet-Weiser, S. (1999). *The most beautiful girl in the world: Beauty pageants and national identity*. Berkeley, CA: University of California Press.

Bates, A. W. (2005). *Emblematic monsters: Unnatural conceptions and deformed births in early modern Europe* (The Wellcome Series in the History of Medicine). Amsterdam: Rodopi.

Benson, S. R. (2005). Hacking the gender binary myth: Recognizing fundamental rights for the intersexed. *Cardozo. Journal of Law and Gender, 12,* 31–80.

Bird, J. (2005). Outside the law: Intersex, medicine and the discourse of rights. *Cardozo. Journal of Law and Gender, 12,* 65–80.

Brady, A. (2011). "Could this women's world champ be a man?": Caster Semenya and the limits of being human. *AntePodium*, 1–16.

Brisson, L. (2002). *Sexual ambivalence: androgyny and hermaphroditism in Graeco-Roman antiquity*. Berkeley: University of California Press.

Brooks, C. (2009, August 25). *Warm welcome home for champ Semenya*. Retrieved March 16, 2016, from http://mg.co.za/article/2009-08-25-warm-welcome-home-for-champ-semenya.

Brown, L. E. C. (2015). Sporting space invaders: Elite bodies in track and field, a South African context. *South African Review of Sociology, 46*(1), 7–24.

Buikema R. (2009). The arena of imaginings: Sarah Bartmann and the ethics of representation. In R. Buikema and I. v.d. Tuin (Eds.), *Doing gender in media, art and culture* (pp. 70–84). New York, London: Routledge.

Butler, J. (1990). *Gender trouble*. New York: Routledge.

Butler, J. (2009, November 20). Wise distinctions. *LRB Blog*. Retrieved March 16, 2016, from www.lrb.co.uk/blog/2009/11/20/judith-butler/wise-distinctions/.

Cahn, S. K. (1994). *Coming on strong: Gender and sexuality in twentieth century women's sport*. New York: The Free Press.

Cahn, S. (2011, January 12). *Testing sex, attributing gender: What Caster Semenya means to women's sports*. Paper presented at the NCAA Scholarly Colloquium, San Antonio, TX.

Carpenter, E. (1921 [1896]). *The intermediate sex. A study of some transitional types of men and women*. London: George Allen Unwin Ltd.

Carty, V. (2005). Textual portrayals of female athletes: Liberation or nuanced forms of patriarchy. *Frontiers: A Journal of Women's Studies, 26*(2), 310–325.

Cohen, H. F. (1994). *The scientific revolution: a historiographical inquiry*. Chicago: University of Chicago Press.

Cole, C. L. (1994). Resisting the cannon: Feminist cultural studies, sport and technologies of the body. In S. Birrell and C. L. Cole (Eds.), *Women, sport and culture*. Champaign, IL: Human Kinetics, 5–30.

Collins, P. H. (1990). *Black feminist thought: Knowledge, consciousness, and the politics of empowerment*. New York: Routledge.

Cooky, C., Dworkin, S. L., & Swarr, A. L. (2013). Justice in sport: The treatment of South African track star Caster Semenya. *Feminist Studies, 39*(1), 40–69.

Cooky, C., Dycus, R., & Dworkin, S. L. (2012). "What makes a woman a woman?" Versus "Our first lady of sport": A comparative analysis of the United States and the South African media coverage of Caster Semenya. *Journal of Sport & Social Issues, 37*(1), 31–56.

Cooky, C., Wachs, F. L., Messner, M., & Dworkin, S. L. (2010). It's not about the game: Don Imus, race, class, gender, and sexuality in contemporary media. *Sociology of Sport Journal, 27*, 139–159.

Darwin, C. (2003 [1859]). *On the origin of species by means of natural selection, or the preservation of favoured races in the struggle for life*. London: John Murray.

Davis, G. (2015). *Contesting intersex: The dubious diagnosis*. New York: NYU Press.

Daston, L., & Park, K. (1995). The hermaphrodite and the orders of nature. Sexual ambiguity in early modern France. *GLQ: A Journal of Lesbian and Gay Studies, 1*(4), 419–438.

Dixon, R. (2009, August 21). Runner Caster Semenya has heard the gender comments all her life. *The Los Angeles Times*. Retrieved March 16, 2016, from http://articles.latimes.com/2009/aug/21/world/fg-south-africa-runner21.

Donoghue, E. (1993). *Passions between women: British lesbian culture, 1668–1801*. London: Scarlet Press.

Downing, L., Morland, I., & Sullivan, N. (2014). *Fuckology: Critical essays on John Money's diagnostic concepts*. Chicago: University of Chicago Press.

Dreger, A. D. (1995). Doubtful sex: The fate of the hermaphrodite in Victorian medicine. *Victorian Studies, 38*(3), 325–370.

Dreger, A. D. (1998). *Hermaphrodites and the medical invention of sex*. London: Harvard University Press.

Dreger, A. D. (Ed.) (1999). *Intersex in the age of ethics*. Hagerstown, Maryland: University Publishing Group.

Dreger, A. D., & Herndon, A. M. (2009). Progress and politics in the intersex rights movement. *GLQ: A Journal of Lesbian & Gay Studies, 15*(2), 199–224.

Eckert, L. (2009). "Diagnosticism": Three cases of medical anthropological research into intersexuality. In M. Holmes (Ed.), *Critical intersex* (pp. 41–72). Farnham: Ashgate.

Epstein, J. (1990). Either/or—neither/both: Sexual ambiguity and the ideology of gender. *Genders, 7*, 99–142.

Fausto-Sterling, A. (2000). *Sexing the body. Gender politics and the construction of sexuality*. New York: Basic Books.

Feder, E., & Karkazis, K. (2008). What's in a name? The controversy over "disorders of sex development." *The Hastings Center Report, 38*(5), 33–36.

Foucault, M. (1977). *Discipline and punish: The birth of the prison*. New York: Vintage Books.

Foucault, M. (1978). *History of sexuality*. London: Penguin Books.

Foucault, M. (Ed.) (1980). *Herculine Barbin: Being the recently discovered memoirs of a nineteenth-century French hermaphrodite*. New York: Pantheon Books.

Geddes, P., & Thomson, J. A. (1889). *The evolution of sex*. London: Walter Scott.

Goldschmidt, R. (1916). Experimental intersexuality and the sex-problem. *The American Naturalist, 50*(600), 705–718.

Goldschmidt, R. (1938). *Physiological genetics*. London, New York: McGraw-Hill.

Grosskurth, P. (1980). *Havelock Ellis: A biography*. New York: Alfred Knopf.

Gruber, N. (2005). Ethics in medicine: With a special focus on the concepts of sex and gender in intersex management. *Cardozo. Journal of Law and Gender, 12*, 117–125.

Hall, S. (1997). The spectacle of the other. In S. Hall (Ed.) *Representation: cultural representations and signifying practices* (pp. 223–279). London: Sage.

Hark, S. (1993). Queer interventionen. *Feministische Studien, 11*(2), 103–109.

Hillman, T. (2008). *Intersex (for lack of a better word)*. San Francisco: Manic D Press.

Hirschfeld, M. (Ed.) (1899–1923). *Jahrbuch für sexuelle Zwischenstufen*. Leipzig: Max Spohr.

Holmes, M. (2008). *Intersex: A perilous difference*. Selinsgrove, Pennsylvania: Susquehanna University Press.

Honegger, C. (1992). *Die Ordnung der Geschlechter. Die Wissenschaft vom Menschen und das Weib. 1750–1850*. Leipzig: Campus Verlag.

Hughes, I. A., Houk, C., Ahmed, S. F., Lee, P. A., & Group, L. C. (2006). Consensus statement on management of intersex disorders. *Archives of Disease in Childhood, 91*(7), 554–563.

Imperato-McGinley, J., Peterson, R. E., Gautier, T., & Sturla, E. (1979). Androgens and the evolution of male-gender identity among male pseudohermaphrodites with 5-alpha reductase deficiency. *New England Journal of Medicine, 300*(22), 1233–1237.

Jones, A. R., & Stallybrass, P. (1993). Fetishizing gender: Constructing the hermaphrodite in Renaissance Europe. In J. Epstein and K. Straub (Eds.), *Body guards: The cultural politics of gender ambiguity* (80–111). New York: Routledge.

Karkazis, K. (2008). *Fixing sex. Intersex, medical authority and lived experience.* Durham, London: Duke University Press.

Karkazis, K., Jordan-Young, R., Davis, G., & Camporesi, S. (2012). Out of bounds? A critique of the new policies on hyperandrogenism in elite female athletes. *The American Journal of Bioethics, 12*(7), 3.

Kessler, S. (1998). *Lessons from the intersexed.* London: Rutgers University Press.

Klebs, E. (1876). *Handbuch der pathologischen Anatomie.* Berlin: Hirschwald.

Klöppel, U. (2010). *XX0XY ungelöst: Die medizinisch-psychologische Problematisierung uneindeutigen Geschlechts und Trans/Formierung der Kategorie Geschlecht von der Zeit der Aufklärung bis in die Gegenwart.* Bielefeld: transcript Verlag.

Knapp, G. (2007, April 10). Women need to raise voices on Imus insult. *The San Francisco Chronicle,* B1.

Krafft-Ebing, R. Von (1903). *Psychopathia sexualis; Mit bes. Berücks. d. conträren Sexualempfindung. Eine klinisch-forensische Studie* (12th edn). Stuttgart: Enke.

Krämer, F. (2007). Die Individualisierung des Hermaphroditen in Medizin und Naturgeschichte des 17. Jahrhunderts. *Berichte zur Wissenschaftsgeschichte 30,* 49–65.

Kristeva, J. (1983). *Powers of horror: An essay on abjection.* New York, NY: University of Columbia Press.

Laqueur, T. (1990). *Making sex. Body and gender from the Greeks to Freud.* London: Harvard University Press.

Levin, D., & Solomon, G. F. (1990). The discursive formation of the body in the history of medicine. *Journal of Medicine and Philosophy, 15*(5), 515–537.

Levy, A. (2009, November 30). Sports, sex, and the case of Caster Semenya. *The New Yorker.* Retrieved March 16, 2016, from www.newyorker.com/magazine/2009/11/30/eitheror.

Long, K. P. (1999). Sexual dissonance: Early modern scientific accounts of hermaphrodites. In P. Platt (Ed.), *Wonders, marvels, and monsters in early modern culture* (pp. 145–163). London: Routledge.

Lorber, J. (1994). *Paradoxes of gender.* New Haven, London: Yale University Press.

Mak, G. (2005). "So we must go behind even what the microscope can reveal." The hermaphrodite's "self" in medical discourse at the start of the twentieth century. *GLQ: A Journal of Lesbian and Gay Studies, 11*(1), 65–94.

McKay, J., & Johnson, H. (2008). Pornographic eroticism and sexual grotesquerie in representation of African-American sportswomen. *Social Identities, 14*(4), 491–504.

Mofokeng, J. (2009). *Millions rejoice for golden girl.* Retrieved March 14, 2016, from www.sowetanlive.co.za/sowetan/archive/2009/08/26/millions-rejoice-for-golden-girl.

Morland, I. (2009). Introduction. Lessons from the octopus. *GLQ: A Journal of Lesbian and Gay Studies, 15*(2), 191–197.

Nederman, C. J., & True, J. (1996). The third sex: The idea of the hermaphrodite in twelfth-century Europe. *Journal of the History of Sexuality, 6*(4), 497–517.

Neugebauer, F. Von (1906). Zusammenstellung der Literatur über Hermaphroditismus beim Menschen. *Jahrbuch der sexuellen Zwischenstufen, 8,* 685–700.

NGBK (2005) *1–0-1 one o' one intersex. Das Zwei-Geschlechter-System als Menschenrechtsverletzung.* Berlin: Neue Gesellschaft fuer Bildende Kunst.

Nye, R. A. (1999). *Sexuality.* Oxford: Oxford University Press.

Nyong'o, T. (2009). The unforgivable transgression of being Caster Semenya. *Bully Bloggers.* Retrieved March 16, 2016, from https://bullybloggers.wordpress.com/2009/09/08/the-unforgivable-transgression-of-being-caster-semenya/.

Paré, A. (1982 [1573]). *On monsters and marvels.* Chicago: Chicago University Press.

Parker, P. (1993). Gender ideology, gender change: The case of Marie Germain. *Critical Inquiry, 19*(2), 337–364.

Platt, P. (Ed.) (1999). *Wonders, marvels, and monsters in early modern culture.* London: Routledge.

Preciado, B. (2003) *Kontrasexuelles manifest.* Berlin: b_books.

Puwar, N. (2004). *Space invaders: race, gender and bodies out of place.* Oxford, New York: Berg.

Reis, E. (2009). *Bodies in doubt: An American history of intersex.* Baltimore: The Johns Hopkins University Press.

Rosario, V. (Ed.) (1997). *Science and homosexualities.* New York: Routledge.

Schleiner, W. (2000). Early modern controversies about the one-sex model. *Renaissance Quarterly, 53*(1), 180–191.

Schultz, J. (2005). Reading the catsuit: Serena Williams and the production of blackness at the 2002 U. S. Open. *Journal of Sport and Social Issues, 29,* 338–357.

Shapin, S. (1998). *The scientific revolution.* Chicago: The University of Chicago Press.

S_he (2003). Performing the gap—Queere Gestalten und geschlechtliche Aneignung. *Arranca* 28, Retrieved March 22, 2016 from http://arranca.org/ausgabe/28/performing-the-gap.

Smith, D. (2009a, August 23). Caster Semenya row: "Who are white people to question the makeup of an African girl? It is racism." *Observer.* Retrieved March 14, 2016, from www.theguardian.com/sport/2009/aug/23/caster-semenya-athletics-gender.

Smith, D. (2009b, September 7). Caster is a cover girl. *Guardian.* Retrieved March 14, 2016, from www.theguardian.com/sport/2009/sep/07/caster-semenya-makeover

Spurgas, A. K. (2009). (Un)Queering identity: The biosocial production of intersex/DSD. In M. Holmes (Ed.), *Critical Intersex* (97–122). Farnham: Ashgate.

Staurowsky, E. J. (2011). A response to "testing sex, attributing gender: What Caster Semenya means to women's sports" by Susan Cahn. *Journal of Intercollegiate Sport, 4,* 54–62.

Terry, J. (1995). Anxious slippages between "us" and "them": A brief history of the scientific search for homosexual bodies. In J. Terry, & J. Urla (Eds.), *Deviant bodies. Critical perspectives on difference in science and popular culture* (pp. 129–169). Bloomington: Indiana University Press.

Terry, J., & J. Urla (Eds.). *Deviant bodies. Critical perspectives on difference in science and popular culture.* Bloomington: Indiana University Press.

Ulrichs, K. H. (1868). *"Memnon." Die Geschlechtsnatur des mannliebenden Urnings. Eine naturwissenschaftliche Darstellung. Körperlich-seelischer Hermaphroditismus. Anima muliebris virili corpore inclusa. 2 Abtheilungen.* Schleiz: Hübscher.

Vannini, A., & Fornssler, B. (2011). Girl, interrupted: Interpreting Semenya's body, gender verification testing, and public discourse. *Cultural Studies ↔ Critical Methodologies, 11*(3), 243–257.

Watson, A. D., Hillsburg, H., & Chambers, L. (2014). Identity politics and global citizenship in elite athletics: Comparing Caster Semenya and Oscar Pistorius. *Journal of Global Citizenship & Equity Education, 4*(1), 1–33.

Wow, look at Caster now! (2009, September 10). *YOU South Africa Magazine,* (114).

Part I
The Clinic

1 Pathologization and Surgery

From 1955 to 1957, the *Bulletin of the Johns Hopkins Hospital* published five articles on so-called hermaphroditic conditions, including an article entitled "Recommendations Concerning Assignment of Sex" by John Money, Joan Hampson, & John Hampson (Volumes 96/6, 97/4, 98/1). This publication series marks the birth of the new treatment paradigm that was installed in what I call intersexualization.[1] The new protocol was called 'sex reassignment' and its goal was to determine the optimum gender of rearing (OGR) for newborns with so-called ambiguous genitalia. The authors wrote that the direction of this 'sex reassignment' procedure should depend on the appearance and functionality of the outer genitalia, considering also the likelihood that they could be surgically altered in accordance with the assigned gender role as either penetrating or penetrable. The aim of the treatment promoted in the research by Money et al. was that 'patients' should "establish their gender with unambiguous certainty" (Money, Hampson, & Hampson, 1955b: 294). Money set the crucial age for this 'sex reassignment' procedure at around 18 months (Money et al., 1955b: 289). The possibility of surgical intervention was central in the determination of the OGR. Money and his team recommended clitor(id)ectomy and vaginoplasty as central to 'intersex management' and their advice is still alive and well today.[2] "The import of the work by Money and the Hampsons in shaping protocols for intersex treatment cannot be overstated," as Lisa Downing, Iain Morland and Nikki Sullivan argue in their comprehensive book on Money and his body of work (2014: 4).

In the course of this chapter, I touch on the discursive and material preconditions that surrounded and impregnated Money's theoretical and practical engagement with what he deemed to be a 'psychosexual emergency.' These preconditions most notably include psychoanalytical referencing and surgical techniques. My intention is to demonstrate how Money et al. constructed their research-object of the psychopathological hermaphrodite. As well, my research reveals that these scientists conveniently ignored the fact that their sample did not support their conclusions. In fact, they construed the concept of gender role in order to render their treatment recommendations comprehensible. Through references to stereotypical masculine and feminine behavior and appearance, they justified the necessity of developing a stable gender role for the intersex child that was congruent with one set of surgically constructed genitals. They were able to essentialize and naturalize a binary notion of gender role by arguing that once ingrained, it is not reversible. I subsequently analyze how the term of innate bisexuality is woven into their neo-Freudian approach to the psychosexual; moreover, I uncover how Money et al. exclusively used those theories by Freud, which fed their argument of the bi-polarity of gender roles, which they based on the theory of dimorphic sexes. In the second part of this chapter, I demonstrate how stereotypes about people who are expected to live in a feminine gender role were used to consolidate the phallocentric and heteronormative organization of twentieth-century society. As I will show, the reinforcement and reinscription of phallogocentrism is literally 'managed' by surgical techniques that erase the ('enlarged') clitoris because it is considered to be phallic flesh that threatens the bi-polar construction of sex and gender on a psychological as well as biological level. I further demonstrate that current debates on intersexualization continue to be heavily influenced by Money's treatment recommendations. Any researcher or clinician who currently takes part in intersexualization refers, if not to Money directly, at least to his collaborators or students.[3]

John Money and the optimum gender of rearing (OGR)

In 1952, John Money wrote his PhD thesis entitled *Hermaphroditism: An Inquiry in the Human Nature of a Paradox* at Harvard University (cited in Karkazis, 2008; Klöppel, 2010). His dissertation sought to demonstrate that anatomy does not determine a person's subjectivity with regards to gender. Most astonishingly, Money later ignored his own findings and started building a new theory around a different, if not to say opposite, paradigm. Throughout his professional life, Money continuously contradicted his own (early) findings in order to establish his treatment paradigm. For infants deemed to have 'ambiguous genitals,' he began advocating surgery within the first 18 months after birth to create genitals that matched the sex assigned to the newborn. This development may in part be due to the academic environment Money joined at Johns Hopkins University. Money himself never studied psychology or medicine, yet he joined a team of medical doctors. Also at Johns Hopkins was Hugh Hampton Young, a genitourinary surgeon who had published on hermaphroditism since 1921 and had extensive experience treating urological conditions (Young and Davis, 1926; Karkazis, 2008: 43). Lawson Wilkins later joined Johns Hopkins as director of the Endocrine Clinic. Wilkins published *The Diagnosis and Treatment of Endocrine Disorders in Childhood and Adolescence* (Wilkins, 1950), which argued that sex reassignment in intersexualized newborns should be decided according to the appearance of the external genitals and not according to the gonads.

Initial sex reassignment should be followed by cosmetic genital surgery and hormones (Wilkins, 1950: 274). At this time, Wilkins still cautioned against radical plastic surgery, but five years later he argued that sex reassignment—including genital surgery—should be pursued in the first 18 months of life (Wilkins, Grumbach, Van Wyk, Shepard, & Papadatos, 1955: 297). His research studied 100 intersexualized people (later also seen by Money, Joan Hampson, and John Hampson). Whereas his main focus was to promote the use of cortisone to conform bodies to match stereotypical secondary sexual characteristics, he also advocated surgery to create artificial genitals 'appropriate' in length, width, and depth. Because gonads (ovaries and testes) were no longer the sole indicator of ones true sex, the scientists also administered chromosomal or hormonal tests, Money, however, often dismissed the results as irrelevant in determining subsequent treatment. The tests were used to define a *true sex* even though the final decision was made according to the presumed *best sex*, the optimum gender of rearing (OGR), which means that for example even though the chromosomal sex was male, the OGR could be determined as female. The publication series from 1955 to 1957 marks the origin of the new treatment paradigm and established Money as the US-American authority on hermaphroditism.

The 1956 paper "Sexual Incongruities and Psychopathology: The Evidence of Human Hermaphroditism" offers an overarching view of the team's approach. This paper was based on 94 people they categorized according to their respective 'psychopathology.'[4] Interestingly, Money, Hampson, & Hampson state: "In 95 per cent of our 94 cases, gender role and orientation corresponded unequivocally with the sex of assignment and rearing" (1956: 43). It is important to note that a year prior to this study, the team had identified seven possible variables when considering hermaphroditic sex and gender:

1 Assigned sex and sex of rearing
2 External genital morphology
3 Internal accessory reproductive structure
4 Hormonal sex and secondary sexual characteristics
5 Gonadal sex
6 Chromosomal sex
7 *Gender role* and *orientation as male or female*, established while growing up
                    (Money, Hampson, & Hampson, 1955a: 302 [my emphasis])

Their findings indicate that variables 1 and 7 correspond (gender role and orientation correspond with assigned sex and sex of rearing) even when they are at odds with 2–6. In other words, those raised as girls, felt themselves to be women and those raised as boys, felt themselves to be men irrespective of their external genitalia, gonads, and/or secondary characteristics.

The Question of Psychopathology

For Money et al., gender role was deeply intertwined with 'sexual orientation' (what today falls under the rubric of 'sexual preference'). To achieve the desired unambiguous gender role, one must necessarily have heterosexual orientation (Money et al., 1955b: 259). Proper gender role and adequate psycho-social integration into heteronormative society (not healthy/functioning ovaries, testes, uterus, etc.) are their preconditions for reproduction. In Money et al.'s 1955 to 1957 publications, the idea of gender role was developed into the key factor for psychosexual development. Psychologically "healthy" and "normal" development depended upon the "inner conviction" of being either male or female (289). Bodily integrity and the preservation of reproductive capacity do not feature in Money et al.'s treatment recommendations. They recommended gender assignment be based on the subject's future ability to become a feminine woman or a masculine man capable of marrying the 'opposite sex' and forming a nuclear family (or adopting if unable to reproduce).

In their assessment of psychopathology Money et al. stated that "in only 94 patients, therefore, was there any question of psychological nonhealthiness on grounds of a demonstrably ambiguous gender role and orientation" (Money et al. 1956: 43). Of the breakdown into "healthy," "mildly nonhealthy," "moderately nonhealthy," and "severely (morbidly) nonhealthy," only five were classified as psychologically "severely (morbidly) nonhealthy" (44). Given the lack of strong correspondence between hermaphroditic individuals and psychopathological traits in their own study, how did Money et al. justify further research into sexual incongruities and psychopathology? What made them continue in the absence of evidence linking "sexual incongruities" to psychopathology? One wonders what the relevance of their research was, given "the most noteworthy finding [...] is the conspicuous absence of severe psychologic disorder" (46)? And one cannot help but note that there was of course a personal motive in this research: the researchers would have been out of a job had there been nothing left to be researched. By continuing to invest in finding a link, they actually in effect created hermaphroditism/intersexuality as pathological.

The Obsession to Categorize

From the list of the seven variables, different so-called intersex conditions were extracted. Michel Foucault has shown how medicine came to be increasingly constituted by "the *medical bipolarity of the normal and the pathological*" in the nineteenth century (1973: 35). Money et al. created new classificatory systems for the concepts of gender, sex, and sexuality. According to Edward Said (1978), the nineteenth century shows an obsession to categorize—Money et al. prove that this obsession continued far into the twentieth century. Lisa Downing's, Iain Morland's, and Nikki Sullivan's book *Fuckology* testifies to Money's "passion for creating taxonomies" (2014: 2). The compulsion to categorize is symptomatic in the processes of intersexualization and all other processes of pathologization. This development of categorization went hand in hand with the establishment of medical expertise and experts authorized to evaluate, diagnose, and treat individuals. So, the production of medical knowledge about the phantasm of intersexuality is intrinsically connected to power.

Foucault discerns the psy-sciences, what he calls the disciplines that are assigned to examine the mind. Bio-power is a set of several disciplinary operations that consists of and encompasses a totalizing means of the organization of subjects (1975). The effect of the organization of subjects is caused by the new disciplinary power he identified; this power is not negative but creative. The power-knowledge complex that is produced over the bodies of intersexualized children excels in the material-discursive effects medical expert knowledge can have. Foucault argues elsewhere that "knowledge is not made for understanding; it is made for cutting" (1977: 154). I suggest that the machinery of diagnosis and treatment in intersexualization evolves literally as a cutting scalpel. In the process of intersexualization this means the development of increasingly intricate criteria, such as the list of seven.

Bio-power is formed at the intersection of scientific knowledge and societal systems of organization. Paul Rabinow and Nikolas Rose "use the term 'biopolitics' to embrace all the specific strategies and contestations over problematizations of collective human vitality, morbidity and mortality; over the forms of knowledge, regimes of authority and practices of intervention that are desirable, legitimate and efficacious" (2006: 197). Thereby they imply that knowledge as well as ethical consideration about life and death, about health and sickness and their demarcations as well as treatments are bound up with political imperatives, decisions, and configurations in the broadest sense. Rose suggests that bio-politics has become the dominant regime of control of bodies "through a system of integrated scientific discourses and social mechanisms" (2001: 38). Foucault uses the terms bio-power and bio-politics to diagnose a social system organized in the body and with the body. The body—'any-body' is subjected to normalizing power/knowledge complexes that are inextricable from medicalizing and pathologizing discourses. Beatriz Preciado calls this an "operation theater": a complex system of coordinates that produces, locates, and organizes bodies in society (2003). It is not only unusual, odd, ambiguous, or deviant bodies that the logic of bio-power extends to, but all bodies and subjectivities in their difference and multiplicities are subjected to this "operation theater." Unique to the case of intersex bodies, however, is that they are subjected to the cutting scalpel as well as to the ideological, symbolical, and political implications of bio-power that penetrate all bodies. Intersexualized individuals are produced in the course of their medicalization and their pathologization; they are created as such. And this is not to say that there wouldn't be numerous variations of bodies—this is to say that there would not only be intersex, female, and male but many more. Male and female bodies are just as much created as intersex—they are just not subjected to surgery. At least not to sex-reassignment surgery—the growing industry of cosmetic surgery of course testifies to the concept of bio-politics. Bio-politics operates in intersexualization in the sense that it organizes sex into only three categories, which are easily count- and therefore controllable.

To support their new treatment recommendations on so-called sexual incongruities and psychopathology, Money et al. called upon a Freudian psychoanalytic discourse of sexuality:

> Few people acquainted with Freudianism and its impact on the theories of psychology and psychiatry can fail to be aware that sexuality, in its broadest

sense, has been given an important place in theories about the genesis of psychiatric disorders. It is appropriate, therefore, to examine the evidence of human hermaphroditism in relation to psychopathology.

(1956: 43)

Ironically, Money et al. write that hermaphroditism must be interrogated in relation to psychopathology—despite the fact that, their own data undermines such a conclusion. Again, in this paper, which is titled "Sexual Incongruities and Psychopatholoy: The Evidence of Human Hermaphroditism" Money et al. state on the first page, that "in 95 per cent of our 94 cases, gender role and orientation corresponded unequivocally with the sex of assignment and rearing" (1956: 43). And they end their classification of psychologically non-/healthy hermaphrodites with the statement that "it proved quite possible for a patient to grow up psychologically healthy in the sex contradicted by chromosomes and gonads and partially contradicted by genial appearance" (52). And, they find that "the surprise is that so many ambiguous looking patients were able to grow up and achieve a rating of psychologic healthiness or of nonhealthiness of only mild degree" (53). Intersexualization, then, becomes only *meaningful* within the discourses that produce it. These discourses are first the production of the very idea of a *true sex*, then the discovery of that true sex and the subsequent surgical and medical construction of a *best sex* (the OGR), all of which rely on the discourse of heterorelational gender roles. Several years of intensive research produced their recommendation to operate in early infancy to secure a stable gender role. This is regardless of the fact that most of their adult patients already claimed stable gender roles—even when their genital appearance was not congruent with their gender role—without such surgery.

Nevertheless, Money and his collaborators argued that early genital surgery was the sole way to guarantee that the intersexualized child would develop into a functioning, non-pathological mono-gendered individual (with a stable gender role) and a mono- and heterosexual body. Accompanying the list of seven variables, they explain that "chromosomal, gonadal, hormonal, and assigned sex, each of them interlinked, have all come under review as indices which may be used to predict an hermaphroditic person's gender—*his* or *her* [original emphasis] outlook, demeanor, and orientation" (Money, 1955: 258). This quotation with the added emphasis reveals the urgency the team felt to assign *his* or *her* gender to the hermaphroditic body, even when the criteria offer no clear sign for an either/or diagnosis.

From the original list of seven diagnostic criteria, the authors wind up excluding gonadal, chromosomal, and hormonal sex as "unreliable prognosticators" of gender role and orientation (Money et al., 1955a: 305). One aspect of their list, however, remains solid: "The most emphatic sign of all is, of course, the appearance of the external genital organs" (Money, Hampson, & Hampson, 1957: 335). What is interesting is that the authors themselves are aware that we are in the sphere of 'prediction.' To recap the process, a hermaphroditic body presents itself to the physician. This body carries both visible and invisible physiological clues that provide critical information revealing its (presumed) tendency towards either maleness/masculinity or femaleness/femininity. In this space of the clinic, the doctors will set their gaze on specific body parts, components, and fluids and produce them as either male or female. In a next step they will impose onto that body a range of future desires and practices understood as exclusively masculine or feminine. Yet, even though the *true sex* of the body is indeterminable in the framework of male and female sex coordinates, a gender role as masculine or feminine is then assigned and its material signifiers (external genitalia) constructed through surgery in order to provide the *best sex*—the OGR. The OGR the optimum gender of rearing—is that which results in a gender role as unambiguously feminine or masculine behavior.

Gender Role, Gender Imprinting, and Language

This gender role is what Money et al. will define and re-define continuously in their publications. The footnote accompanying the diagnostic criteria explains their new concept of gender role:

> By the term, gender role, we mean all those things that a person says or does to disclose himself or herself as having the status of boy or man, girl or woman, respectively. It includes, but is not restricted to sexuality in the sense of eroticism. Gender role is appraised in relation to the following: general mannerism; deportment and demeanor; play preference and recreational interests; spontaneous topics of talk in unprompted conversation and casual comments; content of dreams, daydreams and fantasies; replies to oblique inquiries and projective tests; evidence of erotic practices and, finally, the person's own replies to direct inquiry.
>
> (Money et al., 1955a: 302)

Gender role encompasses, therefore, everything a person thinks, does or feels. Every single move, sensation, and expression can be classified as either masculine or feminine. In their 1957 paper, Money and the Hampsons state that the signs differentiating childhood gender encompass everything from pronoun use to "modes of behaviors, hair cut, dress and personal adornment" (335). Gender role is all-encompassing despite the given that hairstyles, dress, and accessories are always changing according to season and situation, the gender role that derives from them is however considered to be fixed, stable, and constant. Once assigned and surgically constructed, gender role is irreversible in Money et al.'s argumentation. As I will show below, parents play a big role in the stable development of their child's gender role.

Not surprisingly, Money et al. relied upon stereotypical masculine *or* feminine behavior, yet their understanding was reflecting 1950s American society.[5] Moreover, Money et al. rely upon hegemonic white middle-class stereotypes that they assume constitute this white middle-class norm. This is not reflected in their research, and classism and racism are underpinning their research. The results of their research came to be seen as universal. Since Money et al. fostered their theory of adaptability of anybody to these stereotypes disregarding their physiological make-up, they needed to promote the theory that "psychologically, sexuality is undifferentiated at birth and that it becomes differentiated as masculine or feminine in the course of the various experiences of growing up" (Money et al., 1955a: 308). Their move of inaugurating "that sexual behavior and orientation as male or female does not have an innate, instinctive basis" (308). However, is not as flexible as it seems to be:

> The observation that gender role is established in the course of growing up should not lead one to the hasty conclusion that gender role is easily modifiable. Quite the contrary! The evidence from examples of change or reassignment of sex in hermaphroditism, not to be presented here in detail, indicates that gender role becomes not easily established but also indelibly imprinted. Though gender imprinting begins by the first birthday, the critical period is reached by about the age of eighteen months. By the age of two and one-half years, gender role is already well established.
>
> (309, 310)

"Gender imprinting" is here construed as happening even before the acquisition of language, which begins around the age of one to one and a half years. Moreover, the emphasis that "gender imprinting" is irreversible invokes the impossibility to modify gender role. Gender role, in Money et al.'s opinion, since it is so all encompassing, is learning one of two specific behaviors that one practices from the very start of one's life until death. These two sets of behaviors constructed as opposite are located in the coordinate system of a dichotomous and binary sex(ualiz)ed_gendered society. "Gender imprinting" is construed as infiltrating a person's life in every aspect from birth to death—irreversibly and exclusively binary. The analogy of language consolidates this new concept of gender:

> Once ingrained, a person's native language may fall into disuse and be supplanted by another, but it is never entirely eradicated. So also a gender role may be changed or, resembling native bilingualism may be ambiguous, but it may also become so deeply ingrained that not even flagrant contradictions of body functioning and morphology may displace it.
>
> (Money, 1955: 258)

In the first sentence of the above quotation, Money strategically relies upon the ontogenetic learning process of early language acquisition, to present gender role, once imprinted, as durable, indelible, and irreversible.[6] The very next sentence, however, directly states that gender role can be "changed" and by switching back to the language metaphor of "native bilingualism" characterizes both as "ambiguous." Whereas this sentence concludes with the prior conception of gender role as so "ingrained" that it cannot be changed or "displaced," in this middle part of the quotation Money opens the possibility of a twofold imprinting of gender role from the start. In the end, his focus remains on the "deep ingraining" of gender role despite whatever contradictions with body functioning and morphology may later appear. One gender role might be supplanted by the other, but it can never be fully "displaced" or "eradicated" (Money, 1955: 258). Here we can see the conservative, yet plastic concept of gender that Money et al. purported throughout their research. For them, there are only two genders predicted and predicated upon a set of two constructed bodies. Despite their own evidence, however, Money et al. continue to insist upon two exclusive gender roles with two fictional distinct body morphologies. The analogy of language and gender role is used to consolidate the relevance of this new concept and how it can be understood:

> Gender role might be likened to a native language. Once ingrained, a person's native language may fall into disuse and be supplanted by another, but it is never entirely eradicated. So also a gender role may be changed or, resembling native bilingualism may be ambiguous, but it may also become too deeply ingrained that not even flagrant contradictions of body functioning and morphology may displace it.
>
> (Money, 1955: 258)

Gender role is here equated with the acquiring of a native language, which is the use of an argument for a learning theory that can be adapted to other necessary areas of human development. The theory of the imprinting of gender role, its irreversibility, durability, and its indelible character is backed-up by referencing not just academic research but also by calling upon common beliefs about the nature of learning a language. The reference to language acquisition is a rhetorical strategy since it refers to other academic knowledge productions that have been "proven" in terms of an ontogenetic and necessary learning process that is irreversible and permanent (see also Klöppel, 2010). Inconsistently, however, Money acknowledges native bilingualism and therefore in a sense also the possibility of a change of gender role. Nevertheless, this does not lead him to also believe that more than one gender role could be ingrained, even though one can have more than one native language—as in the case of bilingualism. Gender role, one is made to believe, is just as basic for the "individual" to learn, acquire, develop, and stabilize as a native language—bilingualism aside. Even though Money argues that one set of gender roles (either masculinity or femininity) might be supplanted by the other it can never be "entirely eradicated." Interestingly, the concept of gender role is described as being so "deeply ingrained that not even flagrant contradictions of body functioning and morphology may displace it" (Money, 1955: 258). This assertion had earlier been contradicted by Money himself, when he reported of patients who changed their gender role. The distinction between sex and gender and the assumed necessity of concordance between the two according to normative standards (mainly the appearance of genitals) and behavior, thought of as stereotypically feminine or masculine and heterosexual orientation, has been fundamentally contradicted by their sample. However, this does not keep Money et al. from pursuing their agenda of aligning two gender roles with two fictional distinct body morphologies, which are constructed as the mature mono-sexual state of the adult man and woman. Money et al. back up their argument of the necessity of two distinct gender roles based on two separate body morphologies via synechdocal logics.

Synechdocal Logics

Through surgical intervention, Money et al. enabled medicine to erase the visible signs of intersexuality. Predicating their recommendations based on two supposedly distinct body morphologies (maleness/femaleness), they created two corresponding gender roles (masculine/feminine) and disallowed any body or gender role that called this system into question. The surgical intervention they advocated was solely focused on the genitalia. Feminist philosopher Rosi Braidotti argues that "swapping the totality for the parts that compose it, ignoring the fact that each part contains the whole" particularly marks the logic of bio-power (1994: 48). This synechdochal logic blatantly reveals itself in the process of intersexualization. Not whole bodies but specific parts—organs—are organized and classified in a hierarchical scheme. As Riki Wilchins and David Valentine note, "genitals account for only 1 percent of the body's surface area" but they "carry an enormous amount of cultural weight" (1997: 215). Looking at narratives and practices pertaining to intersexualized bodies reveals that genitals ultimately function as the key and the truth of the body. Money et al. assume that

> the external genitals are the sign from which parents and others take their cue in assigning a sexual status to a neonate and in rearing him *[sic!]* thereafter, and the sign above all others, which gives a growing child assuredness of his or her gender.
>
> (1955a: 306)

The *part* seen to be problematic (here: genitals) is modified into a *part* that creates a *whole* (body and fantasy) that comforts the myth that there are only two (opposite, fixed, mutually exclusive, and binary) coordinates of sexed_gendered difference. The genitals are the part upon which surgery is performed, because these parts symbolically organize bodies and identities according to the phallocratic split in society. These parts come to stand for the complete person as individual and are arranged according to a complementarity represented by the phallus and the vagina—the penetrating and the penetrated. They construct a 'mature mono-sexual adult man and woman,' backed up by a selective reading of Freudian psychoanalysis.

From Freudian Bisexual Physical Disposition to Money's Monosexuality

Money et al.'s paper "An Examination of Some Basic Sexual Concepts: The Evidence of Human Hermaphroditism" leans upon Freud's theory of instinctive sexuality. Money et al. understood Freud's theory of innate and constitutional psychic bisexuality as composed of "instinctive masculinity and instinctive femininity [...] present in all members of the human species, but in differing proportions" (1955a: 301). Whereby the terms instinctive masculinity and instinctive femininity appear to be their own coinages. When they pair the terms masculinity and femininity with the term instinctive, they are endowing what we associate with "culture" (i.e., gender role) with a biological "instinct" or essence. Interestingly, Money was the first one to distinguish sex from gender, yet he continuously kept confusing the two—or rather never was clear about how exactly he separated them from each other. And this is also where Money et al. began to misread Freud. They discussed how Freud "construed his theory of innate and constitutional psychic bisexuality on the basis of embryological evidence of an hermaphroditic phase in human embryonic differentiation, and on the basis of anatomical evidence in congenital hermaphroditism itself" (301). Freud indeed implies that the physiology of the human species develops from an originally bisexual (here meaning hermaphroditic) morphology into a 'mono,' that is a 'uni'-sexual morphology, yet, he was cautious about implying an extension to the psychical sphere. Important to note here is, that when Freud speaks about physical bisexuality he means what we understand as hermaphroditic/ intersex and when he speaks about psychical hermaphroditism he means what today we tend to understand as bisexual (meaning, an individual sexually attracted to men and women) and/or homosexual (meaning, an individual sexually attracted to their same sex). Keeping this in mind, I will show how Money et al. misinterpret Freud's very cautious theories about congruencies between physical and psychological development of inversion (homosexuality) and bisexuality (hermaphroditism).

Freud plays with the idea that the physiology of the human species developed from an original, so-called innate bisexual morphology into a mono- or unisexual morphology. In the *Three Essays on the Theory of Sexuality* (1905) he states that "an originally bisexual physical disposition has, in the course of evolution, been modified into a unisexual one, leaving behind only a few traces of the sex that has become atrophied" (Freud, 1961 [1905]: 141).[7] Freud presents hermaphroditism as an originary template for sexual dimorphism that is the difference of appearance between females and males in the human species (see also Salamon, 2004). Humans, therefore for Freud, 'evolve' from a less developed bisexual (hermaphroditic) stage to a more evolved unisexual stage.[8] The historical fact that the terms bisexuality and hermaphroditism have at times been used to name these different, yet interrelated aspects proves their interconnectedness. The very young discipline of sexology at the turn of the century, which Freud also refers to, has had to coin these terms in the first place. The fact that both terms have been changed from the physical to the psychological and the other way around shows that there have always been blurred borders between the body, desire, and subjectivity. Human sexuality is difficult to discern in order to create a consistent taxonomy. Nevertheless, in intersexualization the constant effort to discern male and female is obvious and the frameworks used for differentiation are multiple.

According to Steven Angelides, a biological framework enabled Freud to "preclude[d] the possibility of establishing the independency of psychology" from exactly that biological knowledge production (2001: 63). The evolutionary theories Freud relied on prohibited him from developing a psychology that could exist by itself. Freud, however, was quite cautious with extending this hypothesis of an innate bisexuality of the embryological stage "to the mental sphere" (1961 [1905]: 141). In fact, he states, "it is impossible to demonstrate so close a connection between the hypothetical psychical hermaphroditism and the established anatomical one" (8). For Freud, our "originally physical bisexual disposition" (physical hermaphroditism) does not directly account for "psychical hermaphroditism" (homosexuality). In Freud's reasoning this means that for the proof of homosexuality a physical hermaphroditism has to be present in every homosexual person. Elsewhere, in a footnote in the *Three Essays*, Freud is also dubious of a direct correlation between homosexuality and hermaphroditic physiology: "every individual [...] displays a mixture of the character-traits belonging to his own and to the opposite sex; and he shows a combination of activity and passivity whether or not these last character-traits tally with his biological ones" (86).[9] Freud never saw a causal relationship between an innate or "biological bisexuality" (hermaphroditism) and masculine and feminine personality characteristics.

Freud, as he did with a fair amount of his work, reworked his initial cautious statements about the stance of psychoanalysis towards bisexuality and the meaning of the physical in relation to it. With regard to the concept of innate bisexuality this shift is mirrored in a rephrasing of a sentence in the *Three Essays*. In the editions of 1905 and 1910 he stated that the oppositions of male and female, which are present in bisexuality, can often be replaced by the opposition between active and passive.[10] In the later edition from 1924 this sentence is replaced by one in which the meaning of bisexuality is reduced to the opposition between active and passive in psychoanalysis.[11] In his earlier statement Freud emphasized that he does not know what these gender characteristics are supposed to be, apart from the notions of "activity" and "passivity," which he attributes to men and women, respectively. He states that these character qualities are dependent on biological factors but are also independent from them. In the later version he simply puts active and passive in the place of female and male, respectively—albeit adding that this is a reduction of psychoanalysis. He even goes so far as to say that the significance of this contrast in character-traits is reduced to the contrast between activity and passivity in psychoanalysis.

It is clear, then, that Money et al. misunderstand Freud's concept of innate bisexuality. Whereas they are correct that the founder of psychoanalysis based his ideas on evolutionary theories, Money et al. neglect Freud's skepticism regarding a direct correlation to the mental sphere. By ignoring Freud's doubt about causality, Money et al. also simplify Freud's interesting theory about activity and passivity into a rigid dichotomy of gender roles. In Money et al.'s reading, physiological innate bisexuality (hermaphroditism) is the precondition for psychological innate bisexuality (that is, conceptualizing gender role as exclusively either masculine or feminine). Nevertheless, Freud's evolutionary terms have wide reaching consequences for the processes of intersexualization and the distinction between sex and gender.

The concept of innate bisexuality as the basic and generic stage of human development enabled positioning a so-called hermaphroditic body in an evolutionary framework of physical and psycho-sexual development somewhere on a continuum. This continuum amounts to a construction of the hybrid, which combines, interlocks, and makes intelligible the supposedly two pure entities on two (imagined pure) poles on the continuum and the exclusive framework of reference for any theory of psychosexual development and identity. The evocation of degrees 'in-between' reifies further the two supposedly natural sex_gender poles. 'In-between' is a concept that assumes two entities, whether conceptual, symbolic, semantic, or material. These two entities are located on the two poles of the continuum. This system of coordinates of femaleness/femininity and maleness/masculinity is dependent on its 'in-betweens.' Therefore, 'in-between' could be conceptualized as the origin of meaning in the coordinates of sex and gender. Meaning only appears when two come together; however, these two have to be connected by the hybrid that these entities will form when they merge on the continuum. Argued differently, hybridity can only be constructed if there is an assumed purity of the entities that are merged; moreover, it can always only be merging under a specific criterion that requires definition in the first place. In the bipolarity of such constructions the positions of intersexuality and bisexuality play the significant parts of the other. In their construction as the mediating principle of a particular hybrid they are used to reaffirm the dichotomous and hierarchical categories of the norm. Gilles Deleuze and Félix Guattari have also tackled bisexuality as a conceptual explanation of binary sexes_genders. They state that it is:

> no more adequate to say that each sex contains the other and must develop the opposite pole in itself. Bisexuality is no better a concept than the separateness of the sexes. It is as deplorable to miniaturize, internalize the binary machine as it is to exacerbate it.
>
> (2004: 304)

Bisexuality can only be accepted if it resolves into monosexuality—and it is prone to do so. Bisexuality is by nature bifurcated and can easily be trimmed, differentiated, and dissected into two neat categories. The either/or in intersexualization is prefigured in the concept of bisexuality. As Deleuze and Guattari stress, it is the 'other,' the 'opposite pole' which prohibits bisexuality from being freed from monosexuality; it is the notion of dichotomous complementarity of which bisexuality cannot be deprived. Innate bisexuality is a concept that is inherently problematic since it always implies its own resolution (psychoanalytical) or its evolution (biological) into monosexuality, whether thought of as a bodily original stage of development or as a sexual orientation. The consequences of the misreading of Freud by Money and the Hampsons are particularly evident in Joan Hampson's research, where she advocates clitorectomies in order to guarantee proper feminine gender roles.

Joan Hampson on the Woman with Erectile Phallus

John Money's mentor, the psychologist and physician Joan Hampson moved together with him to Johns Hopkins University in 1951. Her husband John Hampson, also a physician and psychologist worked closely with them, yet only until 1957, when they ended their collaboration, due to unknown reasons (Downing, Morland, & Sullivan, 2014). In the following paragraphs I will look at the only one of Joan Hampson's single-authored publications in the series between 1955 and 1957. Hampson's sample from which she reports in the paper on "Hermaphroditic Genital Appearance, Rearing and Eroticism in Hyperadrenocorticism" (1955) is the same, Money used. The sample grew from 44 patients in 1954 to 94 in 1955 according to the publication series. Joan Hampson reports from her sample of people living as women that "the feminine role had become so thoroughly ingrained that not even a large erectile phallus had challenged the certainty of erotic role" (270). Despite occupying a body with male genitalia, these women were able to disregard the 'call' of their bodies towards masculinity and remain convinced that they are women. Furthermore, Hampson reported that "among the 22 who lived with a contradiction between external genital morphology and assigned sex/gender, 21 had a gender role and erotic practices wholly consistent with assigned sex and rearing" (273), which means that all these people, despite carrying an organ that Hampson classifies as an "erectile phallus," behaved as if they didn't—they behaved as "heterosexual women" (as defined by Hampson) and did not experience instability because of their body's morphology.

One wonders why these people were considered patients in the first place. They all seemed to be at ease with their 'anomaly' that Money et al. were so eager to determine. Furthermore, it is remarkable that Hampson uses the term phallus, since it not just denotes the organ but also the psychoanalytical connotation of the powerful position of the carrier of the phallus in a patriarchal society. Even though the term phallus is the term that is commonly used in medical jargon (instead of the term penis), it testifies to the conceptual framework of a phallocentric and phallocratic practice of viewing bodies and identities. In the proposal for surgical intervention that I will describe below, this not-so-hidden agenda of the symbolic division between men and women exceeds the material division.

In relation to the 'feminine role' and the 'erectile phallus' that Hampson is talking about, the most startling fact for Hampson is that the identification of these women with their so-called male organ has still produced a subjectivity that is at ease with a body morphology that should undermine feminine identification. What is threatened here is not the psychological health or subjectivity of these women but the significance of the penis—the phallus and its relationship to masculinity and all the accompanying behavior as well as the supposedly monolithic relationship between identity and the morphology of the body. I suggest that Hampson and Money subscribed to this monolithic relationship between the phallus and identity, even though their data proved differently. The relationship between gender role and the appearance or possession of specific organs is not necessarily as simple, limited, or binary as they argue. In fact, Hampson sums up her paper by stating that it is not just "possible for an hermaphroditic child to grow into a gender role contradictory of chromosomal, gonadal or hormonal sex" but the child "may also grow into a gender role contradicted by predominant appearance of the external genital organs" (Hampson, 1955: 265). Moreover, Hampson concluded "psychotic symptoms were conspicuous by their absence" in her sample (266). Therefore, the evidence of this research is in no way adequately used but ignored, repressed, put under erasure and into a framework of recommendation for treatment, which feeds the argument for esthetico-functional genital surgery that will be pursued in the Johns Hopkins publication industry.

Surgery Required

Hampson draws the startling conclusion that the "prompt and unequivocating decision of the sex of assignment was found beneficial, along with early reconstructive genital surgery, as required" (273). However, she admits that "reassignment or change of sex in childhood, with or without genital surgery, was found to constitute an extreme psychologic hazard" (273). And she concludes that "physically precocious children presented no untoward social problems of sexual misconduct" (273). The "physically precocious" children she talks about are said not to have any social problems. What is explicitly referred to in the last sentence of the article are "social problems of sexual misconduct," which in light of the atmosphere of the 1950s arguably implies homosexual behavior (or other 'perversions'). What constitutes appropriate sexual behavior is not explained, but there is ample explanation of what guides the researchers in their decision to advocate clitorectomy in early infancy. Hampson states that "the children showed themselves capable of establishing and maintaining serviceable standards of sexual restraint and conduct" (273). It can be assumed that this erotic role, the sexual restraint and conduct that is promoted here for these girls, is heterosexual and passive (penetrateable) and in no case includes the use of the clitoris in any way. The clitoris becomes a dispensable organ in Hampson's paper. "So far as it goes" Hampson states, "the evidence demonstrates that clitoral amputation in childhood or later proved detrimental neither to subsequent erotic responsiveness, nor to capacity for orgasm" (270). "Yet," she admits that "many surgeons have hesitated to deprive a patient of what some authorities have declared the most significant erotic zone in the female" (270). Hampson, nevertheless, also advocates conducting the reassignment procedure (including surgery) in early infancy, assuming that this would not cause "an extreme psychological hazard" (270). What she knowingly withholds is that there is never only one surgical procedure but always several, which will continue into early and late childhood if not into adulthood (see for example Morland, 2004; Karkazis, 2008).

The surgical procedure that Money et al. promoted concerns the genitals and most prominently the clitoris. This clitoris, in a binary framework, is either defined as a macro-clitoris or a micro-penis, according to the genetic testing and the overall diagnosis. Clitoral recession, reduction or amputation became standard protocol through their research. As I argue, a specific normalized embodiment of femininity, masculinity and therefore of heteronormativity underlies the decisions for treatment, and it is assumed, as sociologist Katharina Karkazis states "that intercourse with a surgically constructed vagina would be better than with a small penis" (2008: 116). Feminist biologist, Anne Fausto-Sterling takes this even further and states that in the rationales for surgery "penetration in the absence of pleasure takes precedence over pleasure in the absence of penetration" (Fausto-Sterling, 1995: 131).

'Phallic flesh' and the Erasure of the Clitoris

In the "Recommendations," Money et al. state that a girl with a larger clitoris

> sooner or later, however, [...] comes to realize her oddity. It is preferable that such a child knows, from the time when she can first begin to comprehend it, that she has a clitoris like all other girls, but that it is too big, and will be made smaller surgically.
>
> (1955b: 293, 294)

In the explanatory framework of Money et al. the clitoris undergoes a surgical and discursive erasure. Money et al. state that "girls should also know, incidentally, that whereas boys have a penis, girls have a vagina—in juvenile vocabulary, a baby tunnel—as a double insurance against childish theories of surgical mutilation and maiming" (295). The clitoris is erased as an organ, which can give pleasure via reference to the analogy and complementarity of the penis and vagina. Furthermore, the rhetoric here defeats any argument of mutilation by calling it childish. An organ that is removed from discourse can easily be removed surgically without consideration of leaving a child his_her healthy and perfectly pleasurable organ. Instead, through surgery the 'monstrosity,' the threat to the neatly organized complementary gender roles between men and women, will be erased. Holmes states that this practice testifies to "abuses of power in which silencing and erasing pan-sexual potential are the dominant imperatives" (Holmes, 2000). I argue that this pan-sexual potential which is materially erased in intersexualizing surgery has a homologue in the discursive erasure of the polymorphous perverse in the reception of Freudian psychoanalysis by Money and the Hampsons.

Privileging Vaginal Versus Clitoral Orgasm

In any case, Joan Hampson's research paper reflects a privileging of vaginal orgasm over clitoral orgasm and mirrors Freudian psychoanalysis, which designates clitoral orgasm as infantile and vaginal orgasm as mature. In 1905 Freud 'rediscovered' the clitoris, or rather the clitoral orgasm, actually by inventing its vaginal counterpart. He also defended the notion that the clitoris is a version of the male organ. He states that "man has only one principal sexual zone, only one sexual organ" whereas woman "has two: the vagina, the true female organ, and the clitoris, which is analogous to the male organ" (Freud, 1961 [1931]: 236).[12] As a feminist, one is of course compelled to ask: why not the other way around? And as gender historian Thomas Laqueur states, the clitoris is not "self-evidently a female penis, and it is not self-evidently in opposition to the vagina" either (1990: 234). Freud's era was fanatically trying to justify the social roles of women and men since the first women's movement was just about to threaten the traditional order. According to the era's obsession with the biological make-up of the body and its promising explanatory framework of bio-medical knowledge (predominantly based on Darwinian notions of evolution and development) radical difference was sought. Vagina and penis became to be not just signifiers of sexual difference but also its very foundation. Freud added to this invention of foundational sexual difference when he came up with the theory that only when erotogenic receptiveness has been successfully transferred by a woman from the clitoris to the vagina. She has now successfully developed female maturity and achieved the purposes of her sexual activity (see also Laqueur, 1990: 235).

The Polymorphous Perverse

Freud believed women experience more sexual sensitivity which makes them more corruptible. Since the have two "sex/gender zones" (the vagina and the clitoris) they are more in danger of giving in to their bodies' desire/lust. Women therefore are polymorphous perverse by bodily design. Men supposedly, however, have only one 'sex_gender organ' which makes them more unisexed/gendered.[13] The calculation here meaning: one organ meaning less lust, desire, or sexual excitement. Freud thought that women generally have a greater tendency towards the polymorphous perverse. By the polymorphous perverse, Freud meant undifferentiated possibilities of pleasures (and embodiment). He stated that "under the influence of seduction children can become polymorphously perverse, and can be led into all possible kinds of sexual irregularities. This shows that an aptitude for them is innately present in their disposition"[14] (Freud, 1961 [1905]: 73). This means that every child has the *Anlage* (disposition) to become polymorphously perverse and that the subject eventually learns to contain and control as s_he acquires social, and (I would add), political norms, expectations, censures, rules, and requirements.[15] Freud talks about the child and the "average uncultivated woman" in whom this polymorphous perverse disposition will be contained.[16] However, as Freud states, every woman has the potential to become a prostitute and therefore polymorphously perverse—if not properly cultured into patriarchal, misogynist, heteronormative society. This sexual excess appears when children do not yet know how to concentrate their desire and lust on their genitals, but allow them to roam, as it were, over the entire body. Prostitutes however, according to Freud, deliberately use this so-called infantilism for their job. Freud believed in 1905 that the inclination to the polymorphous perverse is built into the plan of human development, and that a more mature sexuality must be created out of it in the course of cultivation. This implies that class and race play here a big role in suppressing the polymorphous perverse in a specific culture, which is Freud's own moralistic, bourgeois, white, nineteenth-century Viennese culture.[17] Freud abandoned the polymorphous perverse in favor of civilization. This civilization is dependent on clear-cut sexes and genders that have to be aligned to conform to each other via surgical, medical, and psychological treatment—as we can see emblematically in intersexualization.

Genitals and Complementarity

In *The Three Essays* Freud already took care to note that after the transformation of puberty, "the erotogenic zones become subordinated to the primacy of the genital zone" (Freud, 1961 [1905]: 73).[18] Freud gave Hampson and Money the perfect grounds to install the notion that gender is dependent on either male (penile) or female (vaginal) body morphology, and incongruence requires surgical alignment. In the course of time towards intersexualization, the originally polymorphous perverse body and psyche and even the less radical innate bisexuality that Freud declared a disposition, become pressed into two complementary reductive forms of either penis or vagina (1961 [1905]). Laqueur explains, "the social thuggery that takes a polymorphous perverse infant and bullies it into a heterosexual man or woman finds an organic correlative in the body, in the opposition of the sexes and their organs" (1990: 240). The naturalization of heterosexual genital intercourse can only function via a denial of the clitoris in terms of pleasure for the carrier of this organ. Freud's invention of a "dramatic sexual antithesis" results in a "psychologically mature" woman neglecting her clitoridal pleasures and becoming a part of the "perfectly matched couple" of Western society (240). Freud even suggested that the abandonment of the clitoris as a locus of sexual pleasure enhances male desire. He thus again naturalizes heterosexual alliance on which he based reproduction, the family, and indeed Western civilization itself (235). Laqueur describes the history of the clitoris as "a parable of culture, of how the body is forged into a shape valuable to civilization despite, not because of, itself" (236). Elsewhere Laqueur states that the clitoris is the organ of sexual pleasure in women and its "easy responsiveness to touch makes it difficult to domesticate for reproductive, heterosexual intercourse" (1989: 101). That is to say that via the clitoris it is impossible to reinsert "woman into her place in the reproductive economy" (Moore and Clarke 1995: 292). The position of the clitoris (and its responsiveness) destroys the position of women in the economic system—in terms of controllability and functionality. The clitoris has an analogy in the penis—of course always the final arbiter in terms of signification—but maybe it is exactly because of this analogy that the clitoris has to be erased.

In the late Freudian corpus of work, mainly in an *Outline of Psychoanalysis* (1964 [1940]), Freud's earlier trust in the power of psychoanalysis is waning. He starts rejecting his own earlier theories of the polymorphous perversity of body morphology, desire. and the flexible development of the body ego. He sacrifices his earlier claims to the growing powers of biological explanations of the duality of the sexes, which he declares as an ultimate "fact of our knowledge." He surrenders completely: "Psychoanalysis has contributed nothing to clearing up this problem, which clearly falls wholly within the province of biology" (188). Freud laid the basis for Joan Hampson to explain the clitoridal pleasures away from the 'mature' woman. The prioritization of vaginal penetration over the capacity for clitoral stimulation is mirrored in the material effects of intersexualization. Hampson's paper that promotes clitorectomy or clitoridectomy is only a precursor of a body of literature written during the following 50 years that will argue for amputation to deprive infants, children, and adults of their clitoris/phallic flesh for the sake of 'esthetic' genitalia and a 'healthy' psychological development (Braga, Lorenzo, Tatsuo, & Salle, 2006; Gearhart, Burnett, & Owen, 1995; Hendren & Donahoe, 1980).

The Amputation of Healthy Yet Phallic Flesh

A machinery of competing expert recommendations for the best method to amputate healthy genital flesh has operated throughout the second half of the nineteenth century until now. Brooding over the best way to reduce the 'phallic size' of the clitoris and create a 'superior cosmetic result.' In 1995, John Gearhart at Johns Hopkins recommends his own method of clitoral amputation, which is supposed to "preserve the neurovascular bundle in the stump" (Gearhart et al., 1995: 487). Cheryl Chase has responded that she and others who have undergone "clitoral recession" or "clitorectomy" are anorgasmic and that sexual function has been destroyed due to the surgeons "touch" (Chase, 1996). Gearhart replied: "In fact, some women who have never had surgery are anorgasmic" (Gearhart et al., 1995: 487). Another physician admits that "the sexual sensitivity of the clitoris or its remnants may be markedly reduced [after surgery], and some women present with anorgasmia." However, he adds in brackets, "not in all cases may the reason be the surgery, given the prevalence of anorgasmia in non-intersex women" (Meyer-Bahlburg, 1998: 12). I suggest that verbally denying women (intersexualized or not) the ability to orgasm enables these researchers to contain the threat of female sexuality in general.[19] With intersexualized women, there is therefore a literal denial in physical pleasure and orgasm. With non-intersexualized women, the denial is verbal. Instead of looking into remedies for non-intersexualized women who report anorgasmia, they used them to justify the rightness of intersex surgery.

Even though research by Kinsey (1953) and Masters & Johnson (1965a, b) showed that clitorial stimulation was paramount to female sexual pleasure, the clitoris still remains the most under-researched part of the human body.[20] The clitoris is an organ that is considered to be responsible solely for (female) pleasure and because of this exclusive function its investment with politics is immense. I argue, in line with Valerie Traub, that the morphology of the clitoris "is not merely a matter of empirical 'knowledge'; rather, it is constituted by, and includes traces of, desires and anxieties about the meanings, possibilities and prohibitions of female sexuality" (1999: 303). The clitoris has been as much absent from academic and public discourse as the lesbian has; yet, the construction of the clitoris as homologous to lesbian desire shows how anatomy is constructed by gender ideology and heteronormativity. Under-researched but still at the center of attention in intersexualization and threatened by the heteronormative and sexist scalpel of pediatric surgeons, the clitoris becomes the signifier of the conjunction between the (female) body and fears of (female) (homo)sexuality. They are "joined through the imperative of repression" as "the clitoris and the 'lesbian' together signify women's erotic potential for a pleasure outside of masculine control" (302). Reading it through a feminist lens, this line of argumentation implies that the clitoris is invested with the power of the phallus. The clitoris, being associated with phallic flesh and, therefore, phallic power, represents a challenge to the economy of sexual pleasure in a phallocratic society. In a society that is organized around the symbolic organization of a binary split between bodies that carry (men) and those that are the phallus (women), phallic flesh in women, i.e., clitoral flesh, threatens this order. The blurring of these supposed anatomical boundaries makes control according to the clear-cut distinction between identities difficult. The clitoris serves, as Traub puts it "as the authorizing sign of erotic desires, practices and identities" (318). The clitoris, therefore, represents an authority that cannot be granted to women, neither with regards to their erotic potential as being independent from the penetrating penis nor as in regards to their independence from the phallus as the organizing principle of pleasure in general.

This clitoral herstory, as I call it, was initiated by Freud and since then continued by Joan Hampson/John Money and testifies to the complete bi-socialization (via penis vs vagina) of the body's pleasures, desires, and practices. Hampson reiterates the biologisms of her times which are based on an evolutionism that needs two sexes with two very distinct bodily make ups (such as organs, fluids, anatomies, and psychologies). These two sexes enable heterosexual alliance of the all too obviously adapted respective parts through intercourse and subsequent (although not necessary) reproductive activities. All this based on the assumption that heterosexuality is the "natural" state of the "architecture of two incommensurable opposite sexes" (Laqueur, 1990: 233). In intersexualization the clitoris has to be erased to guarantee the absence of phallic flesh in people who are supposed to acquire a gender role as women matching her male counterpart. This Lacanian notion, albeit reduced to the notion of complementarity or sex(ual) difference in general of the ones who *have or carry* the phallus and the ones who *are* the phallus, is not just an imaginary; rather, it is symbolical, cultural, and ultimately political. Phallic flesh in women does not only destroy their prescribed passivity and their symbolic status as the phallus but also the heterosexual complementarity with the penis.[21]

Social Reproduction Versus Biological Reproduction

Interestingly, Money et al. did not consider bodies in terms of *biological* reproductive capacity but rather in terms of *social* reproductive capacity. That is, they determined the probability and likelihood of reproduction according to how well the child might eventually fulfil the gendered stereotypes of 1950s America. According to Money et al. the capacity for biological reproduction alone does not guarantee children:

> Actual childbearing is distinguished from potential biological fertility and is not determined by chromosomal, hormonal, and gonadal sex alone. It is also determined by the social encounters and cultural transactions of mating and marrying, which are inextricably bound up with gender role and erotic orientation.
>
> (1955b: 290)

In other words, you had to be able to *attract* a mate before you *biologically* mate. For them:

> the greater medical wisdom lay in planning for a sterile man to be physically and mentally healthy, and efficient as a human being, than for a probably fertile woman to be physically well but psychologically a misfit and a failure as a woman, a wife, or a mother.
>
> (299)

Interesting here is, that whereas they place greater emphasis on mental health than physical health, they determine mental health according to the subject's adherence to cultural expectations of gender and ability to behave according to heterosexual norms, not according to the subject's own feelings of happiness. A woman who eschews marriage and reproduction or even prefers other women as sexual partners is regarded as a psychological failure. Surgery, then, was performed to prevent the potential of becoming a psychologically misfit woman. In the processes of intersexualization it becomes clear that it is not organs that matter foremost, but their esthetico-functional alignment to the representation of the cultural genitals.[22]

Money et al. inaugurated the surgically modified intersexualized body as the foundation of stereotypical heterosexual subject(ivitie)s. They justified performing medically unnecessary surgery on infants with their belief in the absolute necessity to create bodies and thus subjectivities that conform to heteronormative culture. The so-called 'unfinishedness' of the intersexualized child's body is called upon to convince the parents that the surgery will 'finish' the child into either a girl or a boy. Gender role (and orientation) is construed as exclusively 'either/or.' This either/or needs the support of one specific criterion that Money et al. single out to determine a person's gender as either feminine or masculine. They assume that:

the external genitals are the sign from which parents and others take their cue in assigning a sexual status to a neonate and in rearing him thereafter, and the sign above all others, which gives a growing child assuredness of his or her gender.

(1955a: 306)

They state, however, that "nonetheless, it is possible for a hermaphrodite to establish a gender role fully concordant with assigned sex and rearing, despite a paradoxical appearance of the external genitals" (306). This fact alone suggests that their treatment recommendations were unnecessary in terms of the person's psychological healthiness. Money himself never performed or invented any surgical procedures he just rationalized the use of surgery to create the appearance of binary maleness and femaleness as blueprints for conforming gender roles of masculinity and femininity

Surgical Outcome

In the few contemporary surgical outcome studies of intersex genital surgeries that exist, there is a total lack of consideration of diminished sexual pleasure (e.g., Azziz & Jones, 1990; Azziz, Mulaikal, Migeon, Jones, & Rock, 1986). Recently, however, a number of publications started asking for a reevaluation of the treatment paradigm, mainly arguing with the lack in long-term outcome studies (e.g. Ahmed et al., 2004). Researchers admit that "to our knowledge the long-term effect on sexual function of removing this erectile tissue is unknown" (Baskin, Erol, Li, Liu, & Cunha, 1999: 1018). Others, again, caution that "it will take many more years to prove definitely that the benefits of early operation are maintained throughout puberty" (Jong & Boehmers, 1995: 832). Some researchers, however, still chart the development in surgical procedures such as Baskin et al. who state that "historically the surgical treatment of patients requiring feminizing genitoplasty has evolved from clitoral amputation to clitoral preservation" (1999: 1018). Others have stressed that "the anatomy of the clitoris is not well understood, and studies have shown that most anatomy texts are inaccurate in the size and precise location of this organ" (Crouch, Minto, Laio, Woodhouse, & Creighton, 2004: 137). They reason that "if the clitoris is functionally normal at birth but simply enlarged, it is questionable whether surgery that irretrievably impairs function is acceptable" (138). They realize that women "may resent the destruction of a major part of her sexual potential" (138). In 2003, Minto, Liao, Creighton, Woodhouse, & Ransley composed a survey of 39 women who had clitor(id)ectomy during the last 40 years. They report that only 28 of the 39 were sexually active and all of those had "sexual difficulties." At least there is some consideration of the negative effects, which genital surgeries can have on the intersexualized person; however, the medical establishment is still predominantly advocating surgical intervention in infants (Fox & Thomson, 2005). Peter A. Lee, another contemporary clinician notes that "feedback from former patients generally suggests that considerable variation is preferred to surgery that will compromise sexual sensation. Hence, much greater variation should be more acceptable than previously thought" (2001: 122). Lee, however, still insists that "if genital ambiguity is severe enough to demand a decision concerning sex of rearing, surgical reconstruction must be considered" (122). This case indicates that producers of knowledge in intersexualization desperately want to rescue surgery as a means of treatment. Clitorectomy is still the common practice in intersex surgery, and in Money's words: "Feminizing surgery almost always involves amputation of an enlarged clitoris" (1973: 481). Yet, this is not everything, because "in addition to clitorodectomy, feminizing surgery may require vaginoplasty as well" (481). In the following paragraphs I map out some discourses on vaginoplasty and phalloplasty but focus on the former as it is more frequently applied.

Vaginoplasty

In accordance with Money's treatment protocol most children would be assigned female and are supposed to become feminine—assisted by the practice of clitorectomy and vaginoplasty. Kiira Triea describes Money's gender assignment procedure, which she was subjected to as his patient. Triea was already 14 years old when she first met Money and he asked her if she "wanted to fuck someone or if [she] wanted to be fucked by someone else?" (1999: 142). It is, therefore, assumed that heterosexual (or maybe even homosexual) relationships are based on penetration with one stringently passive (penetrated) and one stringently active (penetrating) part. The criteria for the success of the surgery are still determined by "normal sexual function" or "fully satisfactory intercourse" and do not include sexual enjoyment, "but simply the woman's ability to accommodate a penis without pain or discomfort" (Boyle, Smith, & Liao, 2005: 582). John Gearheart, pediatric urologist, well-known in the field once said: "it's easier to make a hole than build a pole" (quoted in Holmes, 2000: 101).[23] That this is not so 'easy,' however, is shown by the on-going publications that try to enhance vaginoplasty (e.g., Hensle & Reiley, 1998). Moreover, the accompanying practices that are needed to keep open that 'hole' that the surgeons create are immensely intrusive. The practice of dilation goes hand in hand with the construction of a so-called neo-vagina, which mostly has to be carried out by the parents. The parents are responsible for preventing the closure of their newborn's flesh torn apart during the construction of the 'neo-vagina.' The dilation process is assisted by a "Hegar metal dilator of appropriate size," which:

> is given to the family to continue dilatation, usually daily for several weeks, and then progressively less often as postoperative induration subsides. After about 6 mo[nths] have passed, dilatation can be done infrequently (about once a month) to make certain that there is no tendency to narrowing as the child grows older. This seems a small price to pay for being able to carry out most of this reconstructive surgery at a relatively early age, before the child is aware of her abnormality.

> (Hendren & Donahoe, 1980: 753)

Pediatric surgeons Hendren and Donahoe state that "there seems little doubt that parental acceptance of these children is better when their genitalia can be made more normal at an early age" (753). The fact that there are parents who decide to not carry on with the dilation process even though they were told that this would let the tissue grow together is ignored and remains largely unpublished. The great distress that must be felt by the parents, who are literally required to rape their own child daily by forcing this dilator into its body, let alone the child's psychological and emotional well-being and the potentially ruined parent-child relationship, seems of no concern to Hendren and Donahoe. They discuss a "staged approach" that requires more than one operation. These multiple operations cause, however, "a high incidence of vaginal stenosis" (a narrowing of the neo-vagina) (358). They are confident, nevertheless, that the "repair of this anomaly indicates that early one-stage surgical reconstruction can produce a favorable *cosmetic and functional* outcome" (352 [my emphasis],[24] also Jong and Boehmers, 1995: 832). During collaboration between researchers from Canada and Brazil in 2006, Braga et al. do not yet report on the long-term accomplishments but were proud to have achieved "good cosmetic results." For purposes of demonstration they included pictures of children's genitals penetrated by a dilator (Braga et al., 2006: 2199). In any other context pictures of this kind would be deemed child mutilation or child pornography and maybe even rape (for similar practice see Chapter 4 in this volume). These pictures can be described as pornographic if using the definition that porn creates sexual fantasies; these pictures do create the fantasy of sexual 'normalcy' in the doctors.[25] This 'normalcy' is based on the results of a specific esthetic and functional approach that reflects first the erasure of the phallic flesh in women and the construction of the complementary part to the penis, the neo-vagina. The "pleasing cosmetic result" of feminizing surgery is achieved when nothing is left besides a flat area (Bellinger, 1993: 652). This additionally testifies to the argument by Kessler that "lookism" is the final arbiter in the treatment recommendations in intersexualization, which are not only still in place but are constantly reinstalled by some researchers (1998: 109). In another medical paper a side story is told, which gives insight into the importance being placed on a female assigned child having an introitus. Two children died after feminizing surgery due to pneumonia, which they are likely to have caught during their hospitalization (Braga et al., 2006). Karkazis reports a similar case in which a child was treated for heart disease *after* the genitalia were operated on (2008). I suggest that these incidences imply that a surgically and binary sexed_gendered child is more important than the risk to life through surgery.

Treatment Today

Long-term studies are only beginning to be conducted but no research has yet been able to determine the full impact of these surgeries, nor have any of these studies taken a fierce stance towards genital surgery in infants. Understandably, people who have undergone surgeries of such nature are not keen on further contact with the medical establishment or on providing information on their feelings and experiences. Therefore, long-term follow-up research on patients with clitoridectomies and vaginoplasties is very rare. Only since 1998 physicians have little by little begun publishing critical assessments of long-term outcomes of such surgeries (e.g., Berenbaum & Bailey, 2003). Some of them admitted that "aggressive or repeated attempts at vaginoplasty in infancy may be counterproductive" because of extensive scarring (Alizai, Thomas, Batchelor, Liford, & Johnson, 1999: 1590). Furthermore, it has been considered that "dilation is more likely to be successful when undertaken regularly by a motivated young woman who decided to have the surgery than when imposed on an incomprehending child" (Alizai, 1999; Ogilvy-Stuart & Brain, 2004). None of them, however, have thought about why a neonate should need a so-called neo-vagina. One of the first attempts to provide a long-term study in the UK by Sarah Creighton states that it is "still not explained what they mean with sexual functioning, heteronormative bias still present" (Creighton, 2004: 46). Creighton is still one of very few physicians who consider the option of not performing genital surgery in infants.

Normative, 'esthetic' anatomical parameters determine the phallocratic split between an *either* male *or* female 'psychologically healthy' formation of a 'mature gender identity' based on surgically constructed complementary 'holes' and 'poles.' Karkazis states quite adequately that the "clitoris, overwhelmingly tied to female sexual pleasure, is reduced, whereas the vagina, more central to male sexual pleasure, is enlarged" (2008: 154). A so-called micro-penis is therefore enlarged in the processes of intersexualization.[26] The guidelines for phalloplasty that Money found necessary are also still in use. In a prominent publication with Patricia Tucker from 1975, Money stated that

> at the minimum extreme, an erect penis must be something over two and a half inches in length to penetrate far enough into a vagina for a man to begin to feel satisfied with what he can do for his partner.
>
> (56)

Felix Conte and Melvin Grumbach, two surgeons practicing at the Department of Pediatrics, University of California, San Francisco, write no earlier than 2003 that a so called micro-penis in "apparent males" has to be bigger than 2.5 cm and that, similarly, "apparent females" should be investigated who have "clitoromegaly," meaning tissue exceeding the size of 0.9 cm (260). Others have given clear measurements of the normal widths and lengths of the infant's penis (e.g., Feldman & Smith, 1975). Karkazis asks here aptly: "how do you gauge the length of a small, fleshy, spongy organ to an accuracy of 0.2 centimeters?" (2008: 102). The rationales given for the decision for surgery are that "most boys with hypospadias are not able to urinate in the standing position until after surgery, which could keep them from entering the competition with other boys, demonstrating their prowess at urinating at certain distances" (Robertson & Walker, 1975). In 1997 a Dutch research team remarks that "this could play a role in the acquisition of feelings of incompetence, inadequacy, shame, or loss of self-confidence" (Mureau, Froukje, Slijper, Koos, Slob, & Verhulst, 1997: 372). An even more recent paper has asked whether or not the definition of "micropenis" should vary according to ethnicity, which testifies to both the inherent Eurocentrism and the racializing aspects of research on the body and sexuality (Cheng & Chanoine, 2001). These recent publications indicate that the biological functions of the penis are conflated with its social and symbolic ones (Karkazis, 2008: 103). Intersexualization conveyed in phalloplasty materializes the phallocratic organization of society in its purest form. Any possibility of embodiment for a person without a penis as male is foreclosed in this framework. It also assumes that the bodily ego is solely based on sex(ualiz)ed_gendered parameters; additionally, it ignores how surgery also produces embodiment that is arguably extremely more influential in the lives of intersexualized people (Roen, 2009). Phalloplasty also requires repeated surgeries carried out in childhood. A man who has several surgeries which result in the need for life-long catheterization will have a different bodily ego than a man who does not have a catheter and who has not had repeated surgery on his penis. I argue here in line with Iain Morland that "desensitization is not an acceptable side effect of normalizing surgery, because genitalia are for touching, not for looking at" (2009: 296).

Whenever phalloplasty is performed, the question is if an organ can be built that is *big* enough to be a phallus. Size of genitals seems to be an obsession that occurs in very different shapes and intensities. In the nineteenth century, Sarah Baartman, the so-called Hottentot Venus, was represented as sexually excessive because of her visible and prolonged labia (and large buttocks). Her 'monstrous' female genitalia came to be regarded as the signifier for primitive and underdeveloped non-European (here African) sexuality. Even though it is the labia which feature here as the sign for an animal-like and inferior stage in human development, this can easily be adapted to the representation of the clitoris. The civilized, white Euro-American woman is the perfect counterpart for the white, civilized man and therefore cannot have an organ that is pleasurable without being penetrated by the possessor of the phallus. As I discuss, in Chapter 6 in this volume the trope of *development* dominates the discourse of cross-cultural (inter)sexualization and racialization. The development towards maturity in the adult white female here conceptualized as shifting sexual sensitivity from the clitoris to the vagina stands in as a *pars pro toto* narrative. The psycho-sexual erasure and negligence of the clitoris in favor of the vagina is the emblematic process for the material surgical removal of phallic flesh in intersexualized children who are supposed to become wives and mothers in civilized Western culture. The herstory of the clitoris hereby becomes a convincing part of the history of intersexualization and of sex(ual) and racialized/ethnicized difference. Western cultures are based upon genital surgeries of intersexualized people that align two supposedly natural body morphologies to two gender roles upon which the functioning of the split between production and reproduction is built. Intersexualization is normalization in the name of this binary organization of society.

Kenneth Kipnis and Milton Diamond note that thousands of sex reassignment procedures have been performed although "there have been no systematic large-scale studies done to assess the outcome of these procedures" (Kipnis & Diamond, 1999: 178). Most of the cases Money recorded were "lost to follow-up" (179). Money's ideas, as Morland puts it nevertheless "cemented into clinical conventions" (2014: 69)—it was my main concern to show how fundamental inconsistencies in his theories on gender and sex were disregarded and ignored despite the wellbeing of numerous so-called patients being at stake.

Medical Establishment Versus Informed Consent

An accompanying feature of Money et al.'s treatment recommendations mirrors the self conception of the medical establishment as *the* authoritative instance: it has long been medical practice to not inform patients about their conditions. Cancer patients, for example, were, for a long time, not told that they had cancer and that they might die. The authority of this practice has only recently been exposed and criticized by patient self-help groups and their advocacy of informed consent. Respect for autonomy plays a central role in the debate around informed consent (Faden, Beauchamp, & King, 1986). In intersexualization respect for the medically diagnosed person is hardly found. It has become increasingly evident that cases of intersex management do not meet most legal standards of informed consent in the West (Ford, 2001). More recently, ethicist David Hester has examined the exclusion of intersexualized children and their parents from the decision process and the devastating limitations of choice that go hand in hand with this practice (2004). He fundamentally criticizes the concept of informed consent in the process of intersexualization, because for him it is clear that:

> the implicit and fundamental purpose of diagnosis and treatment is *persuasion*, enacted through rhetorical means whereby the body of the patient becomes an object held under a hermeneutics of suspicion, *requiring* (through the assumed premises of argumentation) medical intervention as a necessary conclusion. The result is a circumstance that constrains the decisions of both parents and physicians, such that ethical agency of the patient (or the patient's representatives), even under circumstances of "*informed* consent," can never be granted.

(23)

Clinicians and researchers are still convinced that "surgery in infancy should be aimed principally at creating the appearances of normal external female genitalia, to alleviate parental distress" (Alizai et al., 1999: 1590). This reasoning plays into Money's treatment recommendation of surgery in the first 18 months of life and of not informing the parents of the diagnosis that the expert has conveyed to other medical experts.

Conclusion

The most intriguing fact about the intersexualization process kicked off by Money and his collaborators in the 1950s is that no so-called psychopathologies were exhibited by the very people who ultimately, and sadly, were used to establish one of the most violent medical and psychological practices in Western culture. These practices are based on biological essentialism, gender discrimination, phallocentrism, and heteronormativity. The research Money et al. undertook pursued a phantasm they tried to prove over and over again. Money's researching years and the influence of his work span more than half a century and have been crucial in one of the most terrifying violations of human rights in the West and helped reinstall gender stereotypes that secure the status quo of a heteronormative and sexist society.[27]

Money and researchers standing in his tradition try to find the *true sex* of an intersexualized child to determine the *best sex* in order to decide the optimum gender of rearing. This optimum gender of rearing can either be as a girl or a boy, even though the *true sex* i.e., of the child implies that neither of these two categories follows *naturally* from the sex they determined already—which is intersex. Just as any other newborn is assigned to one of these categories and disciplined into one gender role or another; intersexualized children will experience the same, albeit in a much more violent way. Surgical and medical intervention has become the means by which intersexualized children are made to conform to the norms which through Money et al.'s research have become naturalized. Money et al.'s referencing of Freud's concept of innate bisexuality has provided the grounds for the argument that the intersexualized child is unfinished in its development and therefore has to be surgically 'finished' by the medical expert. The surgery is performed to enable the child to perform one of two heterorelational gender roles according to heterosexist and misogynistic stereotypes. The fact that reproduction is not of central concern but rather the appearance of the external genitals shows that the dichotomous organization of sex(ual) difference exceeds its signification.

Intersexualization and accompanying surgical body modifications—genital surgery on infants who cannot give their consent—are the basis for the constant re-production, reiteration, and enforcement of a heterosexist and misogynistic material and discursive practice. Money et al. construed first a psychopathology where there was none and then argued for surgery which was disguised as the finishing of a body that has not fully developed from an originary embryological state. Second, the surgery, which finished the body away from the underdeveloped will prevent this body to become homosexual. The reasoning behind all this was that a monogendered or unisexual body will be heterosexual.

There has been considerable discussion whether Money was a constructivist or rather deterministic and biologistic in his arguments (Colapinto, 2000; Doell & Longino, 1988; Rogers & Walsh, 1982). It is not my attempt to join this discussion. It is rather my intention to show how Money's and Hampson's desire to find a problem and its subsequent (surgical) solution created drastic and traumatic treatment procedures. Where there might have been a different way of engaging with the multiplicity of human existence—with desires, bodies, and subjectivities—there was a monopolizing practice institutionalized, which cost many people the integrity of their bodies and happy lives. What also becomes clear from my interrogation of Money's research is, as Helen Longino stated "as long as dimorphism remains at the center of discourse, other patterns of difference remain hidden both as possibility and as reality" (1990: 171). Money's research clearly constructs maleness and femaleness as the norm and intersex/hermaphroditic bodies as the aberration that needs to be corrected. It is this medicalizing and pathologizing discursive and material process, which I call intersexualization.

Notes

1. John Money continued his research on hermaphroditism/intersexuality during the second half of the twentieth century. His name entered the public discourse on sex assignment surgery through the so-called John/Joan case reported by John Colapinto in 2000. Money's academic publications include research on most of the so-called pathologies or perversions in sexology

2. The construction of a penis was and is rarely practiced, the reasons for which are discussed below.

3. John Money's collaborators and students are numerous. Most led careers in the general field of sexuality, if not intersexualization directly. Some of those Money co-published with include Anke Ehrhardt, Patricia Tucker, Robert Peterson, Heino Meyer-Bahlburg, and Claude Migeon.

4. Genital surgery is not mentioned, but it is safe to assume that none of the participants had surgery as infants, because this is the practice that Money would be introducing.

5. In 2003 Myra Hird reported from a conference on gender identity which she attended, namely the Tavistock/Portman clinic conference entitled "Atypical Gender Identity Development: Therapeutic Models, Philosophical and Ethical Issues" that took place in 2000. Presentations were given by psychiatrists, psychologists, and physicians working with transgendered and intersexualized people in Britain, Canada, and the USA. Hird reviews most of the lectures and describes the common agreement of most of those on gender identity disorder (GID) in transgendered or intersexualized people that "normal" girls wear skirts and dresses whereas "normal" boys prefer trousers. Hird reports her astonishment about this because "all the female clinicians, including the presenters were wearing trousers. Moreover, none of the female clinicians in the small group wore nail polish or high heels and use of make-up was minimal" (Hird, 2003: 189). All of these criteria were given as expression of "normal" femininity; if these criteria were absent in girls/women, these were considered as indicators for GID. Hird concludes convincingly that the female clinicians "made assessments of GID based on behaviors, roles, clothing, and so on which, by their own assessment, would *render themselves* as suffering from GID" (190). I join Hird in her astonishment and want to stress the fact that anachronistic notions are transported in the knowledge production on intersexualization even by 'professionals' who, would they only look into the mirror, would find themselves disproved in their own limiting and constraining theories.

6. Money also refers here to the theories by Konrad Lorenz who studied instinctive behavior of animals, namely those of wild geese. Trained as a psychologist, Lorenz became interested in evolutionary psychology and conceived the theory that processes of degeneration he recognized in birds may be at work in humans. He was a

member of the NSDAP (Nationalsozialistische Deutsche Arbeiterpartei, i.e, National-Socialist German Workers Party) in Nazi Germany and worked on eugenics and degeneration, which is why his theories were deemed controversial (e.g., Schanse, 2005).

7. German original: "Die Auffassung, die sich aus diesen lange bekannten anatomischen Tatsachen ergibt, ist die einer ursprünglich bisexuellen Veranlagung, die sich im Laufe der Entwicklung bis zur Monosexualität mit geringen Resten des verkümmerten Geschlechts verändert" (Freud, 2000 [1923] vol: V [1905]: 53).

8. The trope that emerges here is 'evolution' from a less developed stage in human development to a more evolved stage, which is the one of a 'unisexual' body. Freud relied on evolutionist theories such as Lamarckian and Darwinian perspectives, and he used them to support his teleological theories of human psychosexual development.

9. German original:

   Jede Einzelperson weist vielmehr eine Vermengung ihrer biologischen Geschlechtscharaktere mit biologischen Zügen des anderen Geschlechts und eine Vereinigung von Aktivität und Passivität auf, sowohl insofern diese psychischen Charakterzüge von den biologischen abhängen als auch insoweit sie unabhängig von ihnen sind.
   (Freud, 2000 [1905] vol. v: 124)

10. German original: "für welchen in der Psychoanalyse häufig der von aktiv und passiv einzusetzen ist" (Freud, 2000 [1905] vol. v: 69). English translation: "A contrast, which often has to be replaced in psycho-analysis by that between activity and passivity" (Freud, 1961: 26).

11. German original: "dessen Bedeutung in der Psychoanalyse auf den Gegensatz von aktiv und passiv reduziert ist" (Freud, 2000 [1905] vol. v: 69). English translation: "a contrast whose significance is reduced in psycho-analysis to that between activity and passivity" (Freud, 1961: 26).

12. German Original: Der Mann hat doch nur eine leitende Geschlechtszone, ein Geschlechtsorgan, während das Weib deren zwei besitzt: die eigentlich weibliche Vagina und die dem männlichen Glied analoge Klitoris" (Freud, 2000 [1931]: vol. V: 277).

13. Freud obviously did not know anything about the prostate, which can be described as a second male sexual organ. The prostate produces part of the semen and is located between the bladder and the rectum. In Western discourse this organ has not been regarded as a sexual erotogenous zone. For thousands of years the prostate is known as a male sexual organ in traditional Chinese medicine or the Tantra.

14. "Es ist lehrreich, dass das Kind unter dem Einfluß der Verführung polymorphous pervers zu werden, zu allen möglichen Überschreitungen verleitet werden kann" (Freud, 2000 vol: V [1905]: 79).

15. The concept of the polymorphous perverse as Freud introduces it is not as unproblematic as I just portrayed it. His first remark of the child being predispositioned to become polymorphous perverse and the prostitute/woman imply that it is a state of 'immorality' in the sense of cultural ignorance but also, and more important for my argument, of immaturity and degeneration. However, I want to caution against a simplifying and careless use of the concept of the polymorphous perverse. In *Eros and Civilization*, Herbert Marcuse argued for the liberation of polymorphous perverse desire and argued for a myriad of ways in which perverse desire and the pleasure principle can collapse the rational ego of repressive Western capitalist institutions. A true liberation of the polymorphous perverse instincts, he argued, can lead to a transformation of so-called sexuality into erotic desire. Marcuse argued that reducing the polymorphous pleasure seeking body to either genital sex or sexuality was a mistake. In fact sexuality "constrained under genital supremacy could never achieve the full eroticization of all the bodily zones that a pregential polymorphous sexuality offered"

(1966: 203). However, Marcuse's championing of the polymorphous perverse can be read as the call for a return to a sort of instinctual truth. But this move would mean positioning the polymorphous perverse in the framework of the "return of the repressed," which is highly problematic from a Foucauldian perspective because it implies a supposedly natural sexual disposition of the human being. To favor the polymorphous perverse, understood in this way, over the phallus in order to liberate the repressed original state of human beings would install an essentialism that can only backfire.

16. "Das Kind verhält sich hierin nicht anders als etwa das unkultivierte Durchschnittsweib, bei dem die nämliche polymorphe perverse Veranlagung erhalten bleibt. (....) Die nämliche polymorphe, also infantile Anlage beutet dann die Dirne für ihre Berufstätigkeit aus..." (Freud, 2000 vol: V [1905]: (97)

17. The fact that Freud's theories derive from a white, bourgeois background reflects in his theories, especially in *Totem and Taboo* and *Civilization and its Discontents* and has immense implications for the racialization process in psychoanalysis. Due to limited space in this chapter I will not be able to elaborate on this but will return later to the processes of racialization in Chapters 4, 5, and 6 in this volume.

18. Gayle Salamon sarcastically notes here that "the same might be said of Freud's own theorizing, which, over time, becomes more consolidated and certain about the categories of sex and genitals" (Salamon, 2004: 102).

19. This is not just one devastating incidence of misogynistic science but emblematic of the ignorance of these researchers operating on genitals that are supposed to become 'female.' Morgan Holmes parallels this logic to telling an abused child that s_he cannot complain about being abused because other children have been abused before (2008: 62). Sara Benson asks accordingly: "what makes doctors think that the child will understand that this is a medically proscribed procedure and not sexual abuse?" (Benson, 2005: 46). Various intersex activists and scholars have reported that children who are subjected to intersex surgery are frightened, shamed, misinformed, and injured and that they experience their treatment as a form of sexual abuse (Triea 1999; David 1995–1996; Alexander 1999).

20. Only recently Helen O'Connell, an Australian urologist, began researching the clitoris. Her research proves that typical textbook descriptions of the clitoris lack detail and are inaccurate. The clitoral anatomy cannot be conveyed in a single one-dimensional diagram because the clitoris is not a flat structure as has been commonly assumed (O'Connell, Hutson, Anderson, & Plenter, 1998); O'Connell, Sanjeevan, & Hutson, 2005).

21. According to Lacan, both sexes identify with the imaginary phallus which paves the way for a relationship with the symbolic phallus, although different for men and women. The man has the symbolic phallus because "he is not without having it" but the woman does not (Lacan, 1989). However, boys first need to give up the imaginary phallus by accepting their castration (in the Freudian sense) to lay claim to the symbolic phallus. The theory also includes that women's lack of the symbolic phallus is simultaneously a kind of possession: they come to *be* the phallus. This misogynistic conceptualization of psychoanalysis has already been criticized powerfully by feminist theorists such as Judith Butler (e.g., 1993, 2000), Jane Gallop (1985) and Theresa Brennan (1993). I apply it here to denote exactly this misogynistic rationale in intersexualization.

22. 'Cultural genitals' is a term that Suzanne Kessler and Wendy McKenna have coined in *Gender: an ethnomethodological approach* in 1978. They used it to indicate that genitals are actually never seen, they are attributed on the grounds of cultural signifiers of gender, such as cloths, habitus, and so on.

23. Morgan Holmes states that:

> Dr. John Gearhart has had this statement attributed to him in print, and I have since heard that he deeply regrets having made it. As appalling a statement as it

was for him to make, it would be a mistake to think that he was either the first or only practitioner to say it, and we may be thankful that he did say it in public, because it illuminates the misogyny informing current clinical practice.

(Holmes, 2008: 148)

24. In the same paper they state that because "the introitus at this time accepted a no. 18 Hegar dilator" they consider their surgery in the child a success. In another case they write that the so-called vagina they created "is a generous size and accepts a 20-mm dilator" (Donahoe & Gustafson, 1994: 358). It is really appalling to read these medical texts that rather come across like two immature boys comparing their penises.
25. I want to thank Robert Davidson for his comment on pornography and the meaning it acquires in this context.
26. The designation of an organ as either a micro-penis or a mega-clitoris is dependent on the sex assignment; both terms denote the same organ.
27. Sarah Creighton, Julie Greenberg, Katrina Roen, and Del LaGrace Volcano remark in their roundtable discussion on "Intersex Practice, Theory, and Activism," which was published in *GLQ: A Journal of Gay and Lesbian Studies* in 2009 that:

the only jurisdiction in the United States to examine this issue carefully is San Francisco. San Francisco issued *A Human Rights Investigation into the Medical "Normalization" of Intersex People*, which declared that the standard medical approach violates the human rights of intersex patients.

(Creighton et al., 2009: 251)

Bibliography

Ahmed, S. F., Morrison, S., & Hughes, I. A. (2004). Intersex and gender assignment; the third way? *Archives of Disease in Childhood, 89*(9), 847–850.

Alexander, T. (1999). Silence = Death. In Dreger, A. D. (Ed.), *Intersex in the age of ethics* (103–109). Hagerstown, Maryland: University Publishing Group.

Alizai, N., Thomas, D. F. M., Batchelor, A. G. G., Liford, R. J., & Johnson, N. (1999). Feminizing genitoplasty for congenital adrenal hyperplasia: what happens at puberty. *Journal of Urology, 161*(5), 1588–1591.

Angelides, S. (2001). *A history of bisexuality*. Chicago, London: The University of Chicago Press.

Azziz, R., & Jones, R. J., Jr. (1990). Androgen-insensitivity syndrome: Long-term results of surgical vaginal creation. *Journal of Gynecologic Surgery, 6*(1), 23–26.

Azziz, R., Mulaikal, R. M., Migeon, C. J., Jones, H. W. Jr., & Rock J. A. (1986). Congenital adrenal hyperplasia: Long-term results following vaginal reconstruction. *Fertility and Sterility, 46*(6), 1011–1014.

Baskin, L., Erol A., Li, Y. W., Liu, W. H., & Cunha, G. R. (1999). Anatomical studies of the human clitoris. *The Journal of Urology, 162*(3), 1015–1020.

Bellinger, M. F. (1993). Subtotal de-epithelialization and partial concealment of the glans clitoris: A modification to improve the cosmetic results of feminizing genitoplasty. *The Journal of Urology, 150*(2), 651–653.

Benson, S. R. (2005). Hacking the gender binary myth: Recognizing fundamental rights for the intersexed. *Cardozo. Journal of Law and Gender, 12*, 31–80.

Berenbaum, S. A., & Bailey, J. M. (2003). Effects on gender identity of prenatal androgens and genital appearance: Evidence from girls with congenital adrenal hyperplasia. *Journal of Clinical Endocrinology and Metabolism, 88*(3), 1102–1106.

Boyle, M. E., Smith, S., & Liao, L. L. (2005). Adult genital surgery for intersex: A solution to what problem? *Journal of Health and Psychology, 10*(4), 573–584.

Braga, L. H., Lorenzo, A. J., Tatsuo, E. S., & Salle, J. L. P. (2006). Prospective evaluation of feminizing genitoplasty using partial urogenital sinus mobilization for Congenital Adrenal Hyperplasia. *The Journal of Urology, 176*(5), 2199–2204.

Braidotti, R. (1994). *Nomadic subjects. Embodiment and sexual difference in contemporary feminist theory*. New York: Columbia University Press.

Brennan, T. (1993). *History after Lacan*. London, New York: Routledge.

Butler, J. (1993). *Bodies that matter: On the discursive limits of 'sex.'* New York: Routledge.

Butler, J. (2000). *Antigone's claim, kinship between life and death*. New York: Columbia University Press.

Chase, C. (1996). Letter to the editor. *The Journal of Urology, 156*(3), 1139–1140.

Cheng, P., & Chanoine, J-P. (2001). Should the definition of micropenis vary according to ethnicity? *Hormone Research, 55*(6), 278–281.

Colapinto, J. (2000). *As nature made him: The boy who was raised as a girl*. New York: Harper Collins.

Conte, F., & Grumbach, M. (2003). Diagnosis and management of ambiguous external genitalia. *The Endocrinologist, 13*(3), 260–268.

Creighton, S. (2004). Long-term outcome of feminization surgery: The London experience. *BJU International, 93*(3), 44–46.

Creighton, S., J. A. Greenberg, K. Roen, & Volcano, D. L. (2009). Intersex practice, theory, and activism. A roundtable discussion. *GLQ: A Journal of Lesbian and Gay Studies, 15*(2), 249–260.

Crouch, N., Minto, Laio, L.-M., Woodhouse, C. R., & Creighton, S. M. (2004). Genital sensation after feminizing genitoplasty for congenital adrenal hyperplasia: A pilot study. *BJU International, 93*(1), 135–138.

David (1995–1996). Clinicians: Look to intersexual adults for guidance. Hermaphrodites with Attitude. *Quarterly Newsletter of the Intersex Society of North America, 7*.

Davis, G., Dewey, J. M., & Murphy, E. L. (2015). Giving sex: Deconstructing intersex and trans medicalization practices. *Gender & Society*, 1–25.

Deleuze, G. & Guatarri, F. (2004). *A thousand plateaus: Capitalism and schizophrenia*. London: Athlone.

Doell, R. G., & Longino, H. E. (1988). Sex hormones and human behavior: A critique of the linear model. *Journal of Homosexuality, 15*(3–4), 55–78.

Donahoe, P., & Gustafson, M. L. (1994). Early one-stage surgical reconstruction of the extremely high vagina in patients with congenital adrenal hyperplasia. *Journal of Pediatric Surgery, 29*(2), 352–358.

Downing, L., Morland, I., & Sullivan, N. (2014). *Fuckology: Critical essays on John Money's diagnostic concepts*. Chicago: University of Chicago Press.

Faden, R., Beauchamp, T. L., & King, N. M. P. (1986). *A History and theory of informed consent*. New York: Oxford University Press.

Fausto-Sterling, A. (1995). How to build a man. In B. W. Berger, S. Watson (Eds.), *Constructing Masculinity* (pp. 127–134). New York: Routledge.

Feldman, K., & Smith, D. W. (1975). Fetal phallic growth and penile standards for newborn male infants. *Journal of Pediatrics, 86*(3), 395–398.

Ford, K. K. (2001). "First, do no harm"—The fiction of legal parental consent to genital-normalizing surgery on intersexed Infants. *Yale Law and Policy Review, 19*(2), 469–488.

Foucault, M. (1973). *The birth of the clinic. An archaeology of medical perception*. London: Travistock Publications.

Foucault, M. (1977). Nietzsche, genealogy, history. In D. Bouchard (Ed.), *Language, counter-memory, practice: Selected essays and interviews* (pp. 139–164). Ithaca, New York: Cornell University Press.

Foucault, M. (2010). *The birth of biopolitics: Lectures at the College De France, 1978*–1979 (Lectures At The College De France) Author: Mi.

Fox, M., & Thomson, M. (2005). Cutting it: Surgical interventions and the sexing of children. *Cardozo. Journal of Law and Gender, 12*(1), 81–97.

Freud, S. (1961 [1905]). Three essays on the theory of sexuality. In James Strachey (Trans. and Ed.), *The standard edition of the complete psychological works of Sigmund Freud* (Vol. 7), (125–245). London: Hogarth.

Freud, S. (2000 [1905]). *Drei Abhandlungen zur Sexualtheorie.* Frankfurt am Main: Fischer Taschenbuch Verlag (Studienausgabe, vol. V).

Freud, S. (2000 [1923]). *Das Ich und das Es* (Studienausgabe, Vol. III). Frankfurt am Main: Fischer Taschenbuch Verlag.

Freud, S. (1961 [1925]). *The ego and the id.* In *The standard edition of the complete psychological works of Sigmund Freud* (Vol. 19), (James Strachey Trans. and Ed.). London: Hogarth.

Freud, S. (2000 [1931]). *Über die weibliche Sexualität.* Frankfurt am Main: Fischer Taschenbuch Verlag (Studienausgabe, Band V).

Freud, S. (1937). Analysis Terminable and Interminable. *International Journal of Psycho-Analysis, 18*(4), 373–405.

Freud, S. (1964 [1940]). Outline of psychoanalysis. In *The standard edition. Edition of the complete psychological works of Sigmund Freud* (Vol. 23), (pp. 141–269), (James Strachey Trans. and Ed.). London: Hogarth.

Gallop, J. (1985). *Reading Lacan.* Ithaca, London: Cornell University Press.

Gearhart, J., Burnett, A., & Owen, J. H. (1995). Measurement of pudendal evoked potentials during feminizing genitoplasty: Technique and applications. *The Journal of Urology, 153*(2), 486–487.

Hampson, J. (1955). Hermaphroditic genital appearance, Rearing and eroticism in hyperadrenocorticism. *Bulletin of the Johns Hopkins Hospital, 96*(6), 265–273.

Hendren, W. H., & Donahoe, P. (1980). Correction of congenital abnormalities of the vagina and perineum. *Journal of Pediatric Surgery, 15*(6), 751–763.

Hensle, T. W., & Reiley, E. A. (1998). Vaginal replacement in children and young adults. *The Journal of Urology, 159*(3), 1035–1038.

Hester, J. D. (2004). Intersex(es) and informed consent: How physicians' rhetoric constrains choice. *Theoretical Medicine, 25*(1), 21–49.

Hird, M. (2003). A typical gender identity conference? Some disturbing reports from the therapeutic front lines. *Feminism and Psychology, 13*(2), 181–199.

Holmes, M. (2000). Queer cut bodies. In J. A. Boone, M. Dupuis, M. Meeker, K. Quimby, C. Sarver, D. Silverman, & R. Weatherstone (Eds.), *Queer frontiers. Millennial geographies, genders, and generations* (pp. 84–110). Madison: University of Wisconsin Press.

Holmes, M. (2008). *Intersex: A perilous difference.* Selinsgrove, Pennsylvania: Susquehanna University Press.

Hughes, J. (1999). *Freudian analysts/feminist issues.* New Haven: Yale University Press.

Jong, T. De, & Bhoemers, T. (1995). Neonatal management of female intersex by clitorovaginoplasty. *The Journal of Urology, 154*(2), 830–832.

Karkazis, K. (2008). *Fixing sex. Intersex, medical authority and lived experience.* Durham, London: Duke University Press.

Kessler, S., & McKenna, W. (1978). *Gender: An ethnomethodological approach.* London: The University of Chicago Press.

Kessler, S. (1998). *Lessons from the Intersexed.* London: Rutgers University Press.

Kinsey, A., Pomeroy, W. B., Martin, C. E., & Gebhard, P. H. (1953). *Sexual behavior in the human female.* Bloomington: Indiana University Press.

Kipnis, K., & Diamond, M. (1999). Pediatric ethics and the surgical assignment of sex. In A. D. Dreger (Ed.), *Intersex in the Age of ethics* (pp. 173–194). Hagerstown, Maryland: University Pub Group.

Klöppel, U. (2010). *XX0XY ungelöst: Die medizinisch-psychologische Problematisierung uneindeutigen Geschlechts und Trans/Formierung der Kategorie Geschlecht von der Zeit der Aufklärung bis in die Gegenwart.* Bielefeld: transcript Verlag.

Lacan, J. (1989). The signification of the phallus. In Lacan, J. *Écrits. A selection* (pp. 311–322). London, New York: Routledge.

Laqueur, T. (1989). Amor Veneris, vel Dulcedo Appeletur. In M. Feher, R. Naddaff, & N. Tazi, (Eds.), *Fragments of a history of the human body* (pp. 90–131). New York: Zone Books.

Laqueur, T. (1990). *Making sex. Body and gender from the Greeks to Freud.* London: Harvard University Press.

Lee, P. A. (2001). Should we change our approach to ambiguous genitalia? *The Endocrinologist, 11,* 118–123.

Longino, H. E. (1990). *Science as social knowledge: Values and objectivity in scientific inquiry.* Princeton, New Jersey: Princeton University Press.

Marcuse, H. (1966). *Eros and civilization: A philosophical inquiry into Freud.* Boston: Beacon Press.

Masters W. H., & Johnson, V. (1965a). The sexual response cycle of the human female: gross anatomic considerations. In John Money (Ed.), *Sex research; New developments* (pp. 53–89). New York: Holt, Rinehart and Winston.

Masters W. H., & Johnson, V. (1965b) The sexual response cycle of the human female: The clitoris: Anatomic and clinical considerations. In John Money (Ed.), *Sex research; New developments* (pp. 90–109). New York: Holt, Rinehart and Winston.

Masters W. H., & Johnson, V. (1966). *Human sexual response.* Toronto, New York: Bantam Books.

Meyer-Bahlburg, H. F. L. (1998). Gender assignment in intersexuality. *Journal of Psychology & Human Sexuality, 10*(2), 1–21.

Minto, C., Liao, L-M, Creighton, S. M. C., Woodhouse, R. J., & Ransley, P. G. (2003). The effect of clitoral surgery on sexual outcome in individuals who have intersex conditions with ambiguous genitalia: a cross-sectional study. *The Lancet, 361,* 1252–1257.

Money, J. (1952). Hermaphroditism: An inquiry into the nature of a human paradox (unpublished doctoral dissertation, Harvard University).

Money, J. (1955). Hermaphroditism, gender and precocity in hyperadrenocorticism: Psychologic findings. *Bulletin of the Johns Hopkins Hospital, 96*(6), 253–264.

Money, J., Hampson, J. G., & Hampson, J. L. (1955a). An examination of some basic sexual concepts: The evidence of human hermaphroditism. *Bulletin of the Johns Hopkins Hospital, 97*(4), 301–319.

Money, J., Hampson, J. G., & Hampson, J. L. (1955b). Hermaphroditism: Recommendations concerning assignment of sex, change of sex, and psychologic management. *Bulletin of the Johns Hopkins Hospital, 97*(4), 284–300.

Money, J., Hampson, J. G., & Hampson, J. L. (1956). Sexual incongruities and psychopathology: The evidence of human hermaphroditism. *Bulletin of the Johns Hopkins Hospital, 98*(1), 43–59.

Money, J., Hampson, J. G., & Hampson, J. L. (1957). Imprinting and the establishment of gender role. *A.M.A. Archives of Neurology and Psychiatry, 77*(3), 333–336.

Money, J. (1973). Hermaphroditism. In A. Ellis, & A. Abarbanel (Eds.), *The encyclopedia of sexual behaviour* (pp. 472–484). New York: Jason Aronson.

Money, J., & Tucker, P. (1975). *Sexual signatures: On being a man or a woman.* Boston: Little Brown & Co.

Moore, L. J., & Clarke, A. E. (1995). Clitoral conventions and transgressions: Graphic representations in anatomy texts 1900–1991. *Feminist Studies, 21*(2), 255–301.

Morland, I. (2009). Between critique and reform: Ways of reading the intersex controversy. In M. Holmes (Ed.), *Critical intersex* (pp. 191–214). Farnham: Ashgate.

Mureau, M., Froukje M. E. Slijper, A. Koos Slob, & Verhulst, F. C. (1997). Psychosocial functioning of children, adolescents, and adults following hypospadias surgery: A comparative study. *Journal of Pediatric Psychology, 22*(3), 371–387.

Nicholson, L. (1994). Interpreting gender. *Signs: Journal of Women in Culture and Society, 20*(1), 79–105.

O'Connell, H., Hutson, J. M., Anderson C. R., & Plenter, R. J. (1998). Anatomical relationship between urethra and clitoris. *The Journal of Urology, 159*(6), 1892–1897.

O'Connell, H., Sanjeevan, K. V., & Hutson, J. M. (2005). Anatomy of the clitoris. *The Journal of Urology, 174*(4), 1189–1195.

Ogilvy-Stuart, A. L., & Brain, C. E. (2004). Early assessment of ambiguous genitalia. *Archives of Disease in Childhood, 89*(5), 401–407.

Preciado, B. (2003). *kontrasexuelles manifest.* Berlin: b_books.

Rabinow, P., & Rose, N. (2006). Biopower Today. *BioSocieties, 1*, 195–217.

Robertson, M., & Walker, D. (1975). Psychological factors in hypospadias repair. *Journal of Urology, 113*(5), 698.

Rogers, L., & Walsh, J. (1982). Shortcomings of the Psychomedical research of John Money and co-workers into sex differences in behavior: Social and political implications. *Sex Roles, 8*(3), 269–281.

Roen, K. (2009). Clinical intervention and embodied subjectivity: Atypically sexed children and their parents. In M. Holmes (Ed.), *Critical* Intersex (pp. 15–40). Farnham: Ashgate.

Rose, N. (2001). Biopolitics in the twenty first century—notes for a research agenda. *Distinktion: Scandinavian Journal of Social Theory, 2*(3), 25–44.

Said, E. (1978). *Orientalism.* New York: Vintage Books.

Salamon, G. (2004). The bodily ego and the contested domain of the material. *Differences, 15*(3), 95–122.

Schanse, A. (2005). *Evolutionäre Erkenntnistheorie und biologische Kulturtheorie. Konrad Lorenz unter Ideologieverdacht.* Würzburg: Königshausen und Neumann.

Traub, V. (1999). The psychomorphology of the clitoris. In S. Hesse-Biber, C. Gilmartin, & R. Lydenberg (Eds.), *Feminist approaches to theory and methodology: an interdisciplinary reader* (pp. 301–329). Oxford: Oxford University Press.

Triea, K. (1999). Power, orgasm, and the psychohormonal research unit. In A. D. Dreger (Ed.), *Intersex in the age of ethics* (pp. 141–144). Hagerstown, Maryland: University Pub Group.

Valentine, D., & Wilchins, R. (1997). One percent on the burn chart: Gender, genitals, and hermaphrodites with attitude. *Social Text, 52*(53), 215–222.

Wilkins, L. (1950). *The diagnosis and treatment of endocrine disorders in childhood and adolescence.* Springfield: Charles C Thomas Publisher.

Wilkins, L., Grumbach, M. M., Van Wyk, J. J., Shepard, T. H., & Papadatos, C. (1955). Hermaphroditism: Classification, diagnosis, selection of sex and treatment. *Pediatrics*, *16*(3), 287–302.

Young, H. H, & Davis, D. M. (1926). Young's practice of urology. Based on a study of 12,500 Cases. *Southern Medical Journal*, *19*(8), 653.

2   Gender Identity Limited

In his lifelong dedication to the scientific study of gender and sexuality, the psychiatrist and psychoanalyst Robert Stoller (1925–1991) wrote nine books, co-authored three and published more than 100 articles. Stoller originally trained at Stanford Medical School, CA and Columbia University, NY. He worked at the Medical School of the University of California, Los Angeles and helped start the UCLA Gender Identity Clinic. Between 1959 and 1990, Stoller conducted numerous case-studies, first on intersexualized and then on transsexual(ized) people. In this chapter I will interrogate ten of his publications between 1959 and 1985 in which he continuously developed the concept of 'core gender identity.' In his first article on "The Intersexed Patient" from 1959, Stoller still used the concept 'sexual identity,' he later replaced it by 'gender identity' and then finally by 'core gender identity.'

During the 30 or so years of his research career, Stoller proposes three components in the development of gender identity: the bodily ego, the parent-infant–relationship, and a biological force. In the course of this chapter I will work my way backwards through these three components starting with the biological force and ending with the bodily ego. Looking closely at how Stoller re-defined and re-worked each of them throughout the years, I interrogate the implications for intersexualization. For example, Stoller first proposed a 'biological force' as being responsible for 'gender identity' but during the years he more or less lost interest in it due to several reasons. In the publications under investigation he also frequently refers to Freudian concepts such as castration anxiety, penis envy, and the Oedipus complex—all of them serve to different degrees and at different times in the development of the concept of 'core gender identity.' These classical psychoanalytical explanatory frameworks are mainly supporting the component of the parent–infant relationship. Major parts of this argument are used by Stoller in determining whether a child is naturally crossing gender boundaries—as in the case of intersex—or pathologically—as in the case of transsexuality. The concept bodily ego will be used to establish a third hermaphroditic gender identity albeit only to dismiss it and reduce gender identity again to the binary.

Stoller's book on *Sex and Gender* from 1968 was crucial for feminism, since the gender-concept has been derived from this book and been introduced to feminist theory by sociologist Ann Oakley in *Sex, Gender and Society* (1972). Stoller, in *Sex and Gender* bases his theories on 85 patients and 63 members of their families whom he either evaluated psychiatrically or subjected them to classical psychoanalysis in the course of the ten years before 1968 (Stoller 1968b: ix). However, the works I will analyze in this chapter are more or less based on not more than 5–10 cases, which he repeatedly draws upon to feed his arguments. Stoller himself describes his approach as having meandered, focusing first on "intersexed, then transsexuals, then those with gender perversions, then the parents of children with gender disorders and, impending, the perversions at large" (xvi). Even though he emphasizes that he focuses on gender identity and processes rather than sexuality, his research is always infused with statements on sexuality, sexual orientation, and preferences (as we call it now).

Stoller's investigation and investment into intersexualization will be questioned mainly by use of current critical work by feminist scholars but also by a close-reading of his publications that is focused on intrinsic contradictions and underlying biases and assumptions. The most interesting feature of Stoller's work for me, however, is one of the byproducts of 'core gender identity:' a hermaphroditic identity. In the course of establishing his theories on a threefold gender identity he also displayed a masculinist agenda trying to prove that masculinity is an achievement and femininity a rather negligible aspect in the development of the human psyche. Whereas also performing a limiting and limited reading of Freudian theories, Stoller tried to accommodate his own findings in a psychoanalytical framework and thereby produced a fundamental worshipping of masculinity. In the course of his research years, Stoller started of with a very cautious stance concerning the treatment of intersex* people—towards the end of his career he echoed Money's emergency paradigm.

From Sexual Identification to Core Gender Identity

In 1959, Stoller co-authored "The Intersexed Patient" with his colleague Alexander Rosen. In this paper, Stoller and Rosen quote J. H. Kiefer's article on "Recent Advances in the Management of the Intersex Patient" from 1957 who defines sex as:

> the overall state of body and mind by which the individual conforms to the masculine or feminine standards of normality in the named sex-determining factors [chromosomes, gonads, hormones, sex organs, and psychic pattern]. It is an algebraic summation of these factors in which no one factor supersedes the others.
>
> (Kiefer quoted in Stoller & Rosen, 1959: 261)

Stoller and Rosen state that they will use the word sex in this sense in their article. Sex is therefore, for them, a matter of conformity to the standards of femininity or masculinity. Interestingly, they do not distinguish between male and masculine or female and feminine. At this point in his career, Stoller still conceptualizes sex as a state of body *and* mind, encompassing both biological features and psychic patterns. Moreover, sex is a formula, an "algebraic summation," much like a mathematical and objective fact. The authors present sex (here especially, the sex of the intersex* patients they studied as something that the medical expert could scientifically determine by adding up the sum total of entities such as organs, body fluids and psychic patterns. Measured against the standards of normality, intersexualization of the patients to be seen meant that the mathematical equation would differ.

The main aim of the paper, however, is to contest the recommendations for sex assignment by Money and the Hampsons (1955b) which state that sex assignment in newborns is best done on the "basis of the external genitalia" and for older children to leave them "in the same sex as that originally assigned" (Stoller & Rosen, 1959: 262). Stoller and Rosen also refer to a study by a D. Cappon, who presented a paper on "Psychosexual Identification (Psychogender) in the Somatic Pseudohermaphrodite" at the Annual Meeting of the American Psychiatric Association in 1958, which contradicts these recommendations by saying that "sex assignment and rearing should always be in the direction of the preponderant somatic sex" (quoted in Stoller & Rosen, 1959: 262). However, Stoller and Rosen did not agree with either version of treatment recommendation. They specifically disagree with the "point of view that all intersexed patients should remain in the identification which started in childhood and persisted into adulthood" (262). Stoller and Rosen state that "the essential criterion [for assigning the intersex* patient's sex] is the strength of the patient's identification with one sex or the other" (262) and that the question of sexual identification needs to be clarified before any plan of treatment (265). Thereby championing the statement of the intersex* person him_herself.

Stoller here still uses the term sexual identification, which he will later replace by gender identity and eventually by core gender identity. At this point, by sexual identification he here means the identification with either the group of men or of women meaning masculinity or femininity. So sexual identification in this communication means what we today—more or less due to Stoller's work—understand as gender identity. Stoller and Rosen report on two cases of "anomalous sexual identification" (262) and conclude in the discussion (here one case representative for the other):

> Nonetheless, one wonders, without adequate evidence, if this patient may not have been propelled, almost against his wishes, by his biological sex. Whether he was compelled to his sexual identifications by unconscious forces of a primarily biological nature or by unconscious forces of a primarily psychic (disturbed identifications in the family) nature has not been determined by our methods.
>
> (1959: 265)

With this statement Stoller sets the tone for his work on gender identity for the next 30 years. The two main aspects he hints at here (1) a biological force and

(2) unconscious forces of a primarily psychic nature (identifications in the family) will feature to different degrees and times throughout his work. However, important to mention is that Stoller and Rosen in 1959 still caution against hasty treatment prescriptions and argue for a careful assessment of sexual identification via psychoanalytical and psychological determination. They still advocate to "assist the patient in determining on his *[sic!]* own to which sex he would like to belong thenceforth" and that "no plan of treatment, in intersexed children or adults, should be embarked on until the question of sexual identification is clarified" (265). This statement clearly contradicts Money et al.'s recommendation and research outcomes, which is focused on the adjustment to the norms of femininity and masculinity. Yet, it also doesn't resonate with the recommendation to assign sex exclusively according to somatic factors.

Three years later, Stoller and Rosen also collaborated with Harold Garfinkel, later known for his ethnomethodological studies.[1] In "The Psychiatric Management of Intersex Patients," (1962) the authors again question Money et al.'s recommendation to operate before the age of 18 months. They doubt the sexual adequacy of surgically altered genitalia and warn against the psychological trauma that might result from repeated surgical procedures. In this paper Stoller, Garfinkel, and Rosen predominantly use the term core sexual identity. And they state that "[a]ssignment of sex is fixed by two general categories, somatic and psychologic" (31). To determine proper sex assignment, in children, they argue that "before the core sexual identity has been established the main problem is to determine the somatic facts" (32). By somatic they mean "(1) Chromosomal sex, (2) gonadal sex, (3) hormonal sex, (4) external and internal genitalia, (5) secondary sex characteristics, (6) body habitus" (31). The preliminary task before determining core sexual identity is therefore the "somatic sex status"—although they do not state why. They elaborate on the second category, which they call the psychologic. The psychologic "is overridingly the most powerful criterion in the development of sexual identity, although usually supported by and exploiting the secondary sex characteristics and appearance of the external genitalia" (31). Stoller and his colleagues here still talk about identification "with one sex or the other" (31) and suggest observing "whether the patient is imitating or caricaturing, or whether true identification is present—that is, comfortable, automatized, nonawkward acceptance of mannerisms, dress, inflections and the like" (31). And they emphasize the "past history" of the patient as regards from which family members "these identifications have been taken, and how easily did these family members fit their own sex roles" (31). They seem to make no difference between sexual identity and sexual identification.

In a 1964 single-authored article called "A Contribution to the Study of Gender Identity," Stoller builds upon his work from 1959, yet revokes his concept of core sexual identity. He now leans toward core gender identity because he understands sexual identity to be ambiguous, "since it may refer to one's sexual activities or fantasies, etc." Core gender identity, instead, "is the sense of knowing to which sex one belongs, that is the awareness 'I am a male' or 'I am a female'" (Stoller, 1964: 220). Stoller now prefers core gender identity because it "clearly refers to one's self-image as regards belonging to a specific sex." By the time of the phallic stage (the timeframe of which is between three and six years), he states, it has already been established. In "normal human beings" it is produced by three components: the anatomy of the external genitalia, the infant–parent relationship, and a postulated biological force (220). For Stoller, the first one is connected to the second one, since "by their 'natural' appearance the external genitalia serve as a sign to parents that the ascription of one sex rather than the other at birth was correct" (223). Moreover, he adds to the first component that "by the production of sensation, the genitalia, primarily from external structures but in the female additionally and dimly from the vagina, contribute to a part of the primitive body ego the sense of self, and the awareness of gender." The infant–parent relationship is:

> made up of the parents' expectations of the child's gender identity, their own gender identity, the child's identification with both sexes, libidinal gratifications and frustrations between child and parents, and the many other psychological aspects of pre-oedipal and oedipal development (223).

Finally, the third component "is the postulated biological force" (223). Beginning at birth when doctors, parents, and midwifes are looking at the baby for the first time and decide if the genitals are rather male or female the course for identification will be set. Hereby several aspects in the parents' own gender identities and their behavior towards the child will influence its (un)ambiguous identification with either the group of men or the group of women. The child's bodily sensations (in the boy more clearly and straight forward than in the girl) will add to this. And finally something biological that "though hidden from conscious and preconscious awareness, nonetheless seems to provide some of the drive energy for gender identity" (220). All these factors work together to produce a *core* in gender identity in Stoller's theory. This *core*, as he states, "remains unchanged throughout life: this is not to say the gender identity is not constantly developing and being modified, but only that at the core the awareness of being either a male or female remains constant" (223). One's self-image as male or female is therefore based on a number of psychological factors (all of them derived from Freudian theory) and the external genitalia, yet also on a so-called biological force.

From Force to Drive—Paving the Way for Freud

In the following years, Stoller hypothesizes this biological force as something that has not yet been—but one day might be—proven scientifically: "some day, such a force may be found to be the algebraic sum of the activities of a number of neuroanatomical centres and hierarchies of neurophysiological functions" (224). Somewhat curiously, he uses the Freudian psychoanalytical term "drive" to discuss the force, despite the fact that psychoanalysis exists outside of the realm of the "endocrinological and neurophysical studies" he suggests that both terms function in a similar fashion. In one case, Stoller reports "*an overpowering drive unalterably and continuously thrusting this child towards maleness*" (1964: 223 [emphasis in original]). The mother of the child has described problems with her little daughter who plays wildly, eats quickly, likes her bicycle, tears her clothes, and so on. She fought with her daughter for years over her failure to act, walk, sit, think, and feel like a feminine girl. "The great effort failed" Stoller concludes, and this was because she "was in fact a chromosomally normal male with a fully erectile tiny penis" (222). After being told that she is actually a boy, she acted as if "she were being told something of which she was dimly aware and had no doubt" (222). After putting on male clothing, s_he:

> has close friends among boys who have no doubt that he is a boy and about whose own masculinity no doubts have been raised. He goes on dates with girls; he is attractive to girls; he is attracted to girls; he has no difficulty in getting dates; he is capable of intense sexual feelings towards girls; he has orgasms with ejaculation either from wet dreams or genital masturbation, in both of which his sexual objects are females (as they were before he was told he was a boy).
>
> (222)

What was formerly a failed girl is now a successful boy. That is, he is successful because he is a *heterosexual* boy who desires, fantasizes about, and dates *heterosexual* girls who, in turn, are attracted to him. The list of things that prove that he is now a proper boy—and has always been, as Stoller asserts—is restricted to his sexual activity and his sexual feeling towards girls, who are his only sexual objects. He is also attracted by and attractive to girls, which Stoller thinks important; it further affirms his masculinity. Moreover, the description of his sexual activity is consolidated by masturbation, ejaculation, and wet dreams. For Stoller these are exclusively male or masculine. The former girl-child was a failure because her behavior was not feminine. Once reassigned to the category of boy, the exact same behavior makes him successfully heterosexual and masculine. The boy's behavior, supposedly driven by an unseen biological force that has always led him to desire girls (even when he was one himself), is now 'normal' because he 'really' is and always has been a heterosexual boy. Once *he* is allowed to wear male clothing, *she* is no longer a threat and *his* true essence emerges.

In the same paper, Stoller asks himself and his reader if it is really necessary to invoke a biological force to explain the data on a child without penis and scrotum (which produce genital sensation—leading to a bodily ego) who defied the parental attitude of raising it as a girl. The child was convinced that it was a boy. Stoller's interest is in how to tell that this is "not a child, like many others, who, in a pathological relationship with its mother, has developed a very masculinized gender identity? The word abounds in 'butch' homosexual woman" (224). Stoller contemplates that there are "parental attitudes" that can produce "homosexuality and other perversions" yet, they do not cause an intact core gender identity that is in contradiction to the appearance of the genitals (224). In search for his postulated/un- or preconscious biological force, Stoller patholo-gizes lesbian desire embodied by the butch. He reports that the "calm, sure mas-culinity of the child [...] shows itself in glaring contrast to the 'butch'" (224). The 'abnormal' butch can now be a normal boy, established on the grounds of sexual orientation. Stoller states that the child "always felt (though not con-sciously) that he was a male. He did not shift from female to male, but only had the rights of maleness confirmed by society" (223). These rights of maleness seem to be entailed in the biological force. The natural confirms the normal—here especially pure masculinity that is not contaminated by femininity or lesbian desire. It has been there all along—the natural calling toward masculinity by a supposedly male biological force. Stoller refers back to his definition of gender identity as something that is intrinsically interwoven in the person and not easily shifted. As already stated above, Stoller defines core gender identity as meaning an 'unalterable sense' of being either male or female. Hereby, gender identity is not only essentialized and naturalized but masculinity and femininity are positioned as the only feasible (and 'promising') possibilities for the devel-opment of identity. Although intersex* is positioned as the natural cause—via the biological force—and thereby construed as the natural, it is also construed as that which will produce the normal: heterosexual desire and corresponding gender performance—if adjusted to the male gender role.

One of my main interests, however, in Stoller's body of work, is his reference to Freudian psychoanalysis. As I have already quoted above, he reports "*an overpowering drive unalterably and continuously thrusting this child towards maleness*" (1964: 223 [emphasis in original]). The biological force "though hidden from conscious and preconscious awareness, nonetheless seems to provide some of the drive energy for gender identity" (220). And without pause Stoller goes on to refer to Freud (1905 and 1937) who, according to Stoller "never abandoned the position that biological forces were an essential though unmeasurable part of personality development" (220). Stoller states that Freud was not able to investigate these biological forces further since he did not have the adequate clinical material. Stoller, however, attempts to address this issue again and uses his own clinical examinations to "gain insight into these forces" (220). It is the use of Freudian terminology and concepts, which will lay the grounds for Stoller to develop the concept of a third core gender identity—the hermaphroditic gender identity. And it will lead him to recommend intersex treatment until the age of two or three the latest—hereby echoing Money and the Hampsons (1955b) rather than cautioning against it as he used to do before (Stoller & Rosen, 1959). He will make one last attempt to rescue the hidden/ postulated/unseen/silent biological force in 1968 in an article called "Can a Bio-logical Force Contribute to Gender Identity." But his Freudian legacies will overpower the endeavor, and he will simplify multisided identificatory processes into core gender identities.

Freud introduced the sexual drive (Trieb) in the *Three Essays* (1905), which is the account Stoller refers to. For Freud, the sexual drive (Trieb) was the expression of innate sexual tendencies in all humans from the day we are born.

The drive (originally translated as 'instinct' by James Strachey) was fundamental for Freud's thinking. He assumed it to be biological and innate, yet, as producing libido and sexual excitement rather than maleness and femaleness.

By an "instinct" is provisionally to be understood the psychical representative of an endosomatic, continuously flowing source of stimulation, as contrasted with a "stimulus," which is set up by single excitations coming from without. The concept of instinct is thus one of those lying on the frontier between the mental and the physical. [...] The source of an instinct is a process of excitation occurring in an organ and the immediate aim of the instinct lies in the removal of this organic stimulus.

(Freud, 1961 [1905]: 168)[2]

The Freudian understanding of drive (Trieb) is clearly confined to excitement caused by an organ, which seeks dissolution of the cause for excitement. It had absolutely no relationship to Stoller's invention of an inner force that 'drives' a person toward a sense of self as male or female. By saying that the drive thrusts the child towards maleness, Stoller intrinsically connects desire with identification. The biological force Stoller implies—with reference to Freud shifts the original Freudian conceptualization of drive/instinct. By implying the Freudian drive when talking about identification as male or female, Stoller implies heterosexual orientation in an identification process. Stoller assumes therefore that the drive causes and is caused by a heterosexual orientation/preference, thereby misreading Freud, who was careful enough to not talk about the object of desire, when talking about drive/instinct. Stoller of course mirrors the atmosphere of his times yet also he re- and confirms homophobic undertones. I suggest that this use of drive as the reference point for the postulated biological force represents the impossibility of separating sex from gender without implying sexuality (or sexual orientation/preference) within the heteronormative framework of Stoller's work. Consequently, Stoller's theories of core gender identity are intrinsically tied to a heteronormative bias according to which this assumed biological force or drive has to be discerned. Stoller's concept of core gender identity as innate (implying Freud's drive theory and the hypothesized biological force), naturalizes gender identity in terms of feeling to belong to one or the other 'natural sex category' that naturally *and* normally desires the 'opposite natural sex category.'

The second case Stoller refers to in this article is that of Agnes, whose story is used to feed the biological force argument. It will however, change the course of Stoller's arguments on the biological force and shift his entire research from intersexuality to transsexuality. Agnes was raised as a boy but at 17 s_he "developed *all* the secondary sex characteristics of a girl" (Stoller, 1964: 225 [emphasis in original]). Furthermore, Stoller reports that "pathological examination of the testes [after surgical removal] revealed them to be the source of large amounts of oestrogen produced since puberty when the feminine appearance developed" (225). Stoller sees his biological force proven for the second time, since in this case the "core gender identity was female, despite the fact that the child was an apparently normal-appearing boy and was also genetically male" (225). This case will gain greater significance in the course of this argument since Agnes, in fact, started taking hormones (oestrogens) at the onset of puberty and hence was not subjected to a supposed foetal (or natural) biological force. However, this was only later revealed to Stoller (1968a)—causing him to abandon the biological force and making him add a new dimension to core gender identity—the pathological pre-oedipal *trias* with the tomboyish mother who cannot let go of her son so that he can become properly masculine. I analyze this aspect, attributed by Stoller to the infant–parent relationship, in the following paragraphs. First, however, I want to map out Stoller's stance towards the concept of gender, since it will also be part of Stoller's reworking of the Oedipus complex.

Supporting Money's Recommendations

One year later, in 1965, Stoller publishes "Passing and the Continuum of Gender Identity" in which he repeats that "the term 'gender' connotes psychological aspects of behaviour related to masculinity and femininity. It does not have the same meaning as 'sex'" (197). Sex, however, remains unexplained. The main task that Stoller pursues in this paper is to distinguish gender role, which is Money's (1955a) coinage, from gender identity, which is his own creation. Gender role according to Stoller is a purely culturally determined role and can shift, whereas core gender identity remains constant. Stoller adds: "A clear-cut gender identity leads to a habitually clear-cut confrontation of society that is called 'gender role'" (198). Moreover, he states, "it is possible to shift gender role, even while gender identity is constant" and concludes: "one plays a role, but possesses an identity" (198). Stoller however contradicts Money in calling this conviction identity rather than role as Money and the Hampsons do. He however echoes Money and the Hampsons (1955, 1956, 1957; and Hampson 1955) in saying that "attempts to reverse the child's sex and gender are increasingly less successful as the child grows beyond this relatively critical period" (199). This is the point at which gender identity plays into the treatment recommendations by Money and the Hampsons. Stoller states that people who are uncertain to which gender they belong—a shaky gender identity–can shift gender role easily. He suggests that "the less well established the gender identity, the more intense the cross-gender impulses" (201). As concerning the argument for early surgery by Money and the Hampsons (1955b) this confirms their recommendations as Stoller now dates the establishment of core gender identity before the third birthday. Every attempt "to reverse" it is deemed to be unsuccessful. He also mentions something like a hermaphroditic identity for the first time. He reports a "rare" gender identity that has been developed by "a hermaphroditic male brought up as a girl." The person is "uncertain if he is either male or female. This uncertainty is as much a part of his identity as certainty is a part of the more normal identity" (198). Stoller here already sets the tone for his definition of a third core gender identity, the hermaphroditic gender identity, which will be used to consolidate the normalcy of core feminine and masculine gender identities.

The Parent–Infant Relationship

In "A Further Contribution to the Study of Gender Identity" in the *International Journal of Psycho-Analysis* (1968a) Stoller revisited the Agnes case.[3] I consider this a fundamental shift in his research because from now on he intensified his focus on the tomboyish mother (a term coined by Stoller). This focus is supported by Stoller's development of the second aspect in core gender identity: the infant–parent relationship, or rather the mother–infant relationship.

Stoller explains how Agnes confessed to him, that "she had not become feminized as the result of oestrogens produced in her testes but had rather been taking oestrogens since puberty" (Stoller, 1968a: 365). "Thus," Stoller has to admit, "she could not have been feminized as a result of a 'biological force' " as he had wrongly reported:

> nor could the development of her secondary sex characteristics be taken as evidence of a biological force that, starting from early childhood on, had so influenced the development of her gender identity that she had felt herself to be really a female.

(365)

Agnes, so Stoller reports, was instead a transsexual, "a biologically normal male, who nonetheless feels himself to be a female" (366). Stoller now re-defines Agnes as transsexual, whereas before he knew that she induced her outer appearance by the intake of oestrogens he defined her as intersex. Agnes cannot be intersex anymore, in his reasoning, because it was not her body, with which the pre-natal biological force called her towards femininity, but it was a self-induced hormonal change in the body appearance. It is now the behavior of Agnes that Stoller sees in need of explanation, albeit not through the biological force anymore but through a psychoanalytical framework. The shift caused by this case is reflected in his entire body of work. Stoller will from now on concentrate on a re-working of Freudian concepts in order to establish a firm notion of normal male or female gender identities.

Intersexualization and Transsexualization

As becomes clear, intersexualization is intrinsically connected to transsexualization. Both processes are based on the newly developed gender-concept, that is, the distinction of sex and gender. Whereas for Stoller intersexuality is biologically induced, transsexuality is a psychological pathology, which is produced by the failure of the Oedipal conflict. In the following paragraphs, I focus on his pathologization of transsexuality i.e., transsexualization, which he notwithstanding interconnects with intersexualization on the level of the pre-oedipal. I argue that Stoller, in his quest to rehabilitate his hypothesis of an essential feature of binary sex in the body, i.e., the postulated biological force disproven by Agnes, turns to the psychoanalytical concept of the Oedipus complex in order to induce an equally essential feature in the psyche. Since the publication of *The Interpretation of Dreams* (1896) the Oedipus complex is to be found in nearly all of Freud's theories as the unquestioned basis of culture. The Oedipus complex is foundational for the theories in *Totem and Taboo* (1931), *Culture and its Discontents* (1927), the *Three Essays on Sexuality* (1905), the concept of the unconscious, as well as the distinction between the ego and the id and the place of the phallus.[4] Freud's theory of psychosexual development is separated into the oral (0–1 years), anal (1–2 years), phallic (3–6 years), latency (childhood), and genital (puberty) stage. These stages, as Freud identified them, are non-negotiable: everybody has to go through them. Yet, during his research Freud frequently changed the timeframes in which these stages take place. He acknowledged that development varies between individuals and that different stages can even exist simultaneously within a given individual. In the phallic stage the passing of the Oedipus complex or the resolution of the Oedipal conflict takes place. The Oedipus complex is derived from the myth of *Oedipus Rex,* who married his mother and killed his father (albeit unknowingly). Freud uses the myth to describe how the (male) child desires his mother and develops anger towards his father, because the father is the one who can have the mother. The threat the father poses to the child produces castration anxiety in the boy. The language often used to describe this and to universalize it is the one of 'opposite' and 'same' sex parents who are desired or envied (hated). C. G. Jung has used the term *Elektra complex* to describe the process female children go through.[5] Freud, however, called it 'feminine Oedipus attitude.' Even though Freud could never sufficiently explain psychosexual development in girls he described the phallic stage through the clitoris, the 'penis-equivalent,' in which girls experience penis-envy and finally accept that they cannot have a penis. For women he defined an extra stage in which they transfer the importance and sensitivity of the clitoris to the vagina (see also Chapter 1). The resolution of the Oedipus complex occurs by identification with the parent of the 'same sex' and by the repression of the sexual interest in the parent of the 'opposite sex.' It is here where the male child learns to identify with the 'same sex' and to shift the desire for the mother to other representatives of the 'opposite sex.'[6] For girls this means that they have to learn to identify with the formerly loved object of the mother and have to shift their desire to the father. Freud, however, acknowledged that the Oedipus complex process is not as smooth as that, and hence introduces bisexuality as a complicating factor in the Oedipus complex. Bisexuality becomes a fundamental concept in the Oedipus complex because it can describe desire as well as identification. Freud introduced the notion of bisexuality as an orientation of desire in relation to the Oedipal conflict and called it the negative Oedipus complex:

> Bisexuality [is] originally present in children: that is to say, a boy has not merely an ambivalent attitude towards his father and an affectionate

object-choice towards his mother, but at the same time he also behaves like a girl and displays an affectionate feminine attitude to his father and a corresponding jealousy and hostility towards his mother. It is this complicating element introduced by bisexuality that makes it so difficult to obtain a clear view of the facts in connection with the earliest object-choices and identification, and still more difficult to describe them intelligibly.

(1961 [1925]: 33)

Freud's ambivalence and caution towards his own concepts and theories is apparent. Freud states that bisexuality plays a role in identification *and* object-choice and thereby opens the space of the Oedipus complex for a much more flexible theory of identification in relation to desire. He admits that he never gained a 'clear view' of the workings of bisexuality.

Bisexuality and the Bifurcation of Gender

Judith Butler explains that bisexuality features here because Freud could not provide a place for homosexuality in the healthy development of the child. Judith Butler therefore claims that "bisexuality is the coincidence of two heterosexual desires within a single psyche" (Butler, 1990: 77). This testifies to the absurdity of restraining either—identification and desire—to a binary system that is based on the heterosexual matrix. Steven Angelides sums this up as "to identify is to repress bisexual desire, and to desire it is to repress bisexual identity; yet to identify and to desire is to be predisposed bisexually" (2001: 63). It is this distinction between 'same sex' and 'opposite sex' which makes a non-heteronormative interpretation of Freud difficult.[7] This distinction presupposes that the child internalizes either a masculine or feminine subject position, which happens via a heterosexual desire for the 'opposite sex' parent. The child then identifies with the other, the 'same sex' parent by converting the desire it felt before. The resolution of the Oedipal conflict went wrong if the (male) child did not give up the mother in order to identify with the father. Therefore, remaining in the logic of the Oedipus complex, Freud had to explain homosexuality as a neurosis. The Oedipus complex is, therefore, intrinsically heteronormative and cannot account for anything outside this complementary economy of the nuclear family.

Stoller uncritically takes on Freud's heteronormative bias and neglects the workings of the negative Oedipus complex. Stoller, in fact, goes even further than Freud and claims that homosexuality is a perversion (Stoller, 1964: 224). In reference to the Oedipus complex, but reifying it on the basis of the infant–parent relationship and the failure of the mother to let the son separate from her body, Stoller gives four reasons for the Agnes case. In most of his research, Stoller focused on boys and developed his theories without paying attention to girls. If Stoller had interrogated the development of girls, his theories would probably have taken a different route. In the framework of the Oedipal trias, however, there is no possibility to account for transsexuality in girls because gender roles are clear-cut and the positions of the father and the mother non negotiable, whatsoever. The four reasons for transsexuality in boys, based on the case of Agnes and one other patient are as follows: First, the mothers are tomboyish and bisexual in Stoller's terms, meaning that they are not properly feminine (wearing suits instead of dresses etc.). Second, the fathers are passive men, themselves effeminate and almost completely physically absent from their families. Third, both parents would permit the child's effeminacy to develop. And, fourth, the mother would have "astonishingly excessive physical contact" resulting in the child "acting in a feminine manner and showing his desire to be a female" (Stoller, 1968a: 366). Stoller concludes his study on transsexuality by stating that "overwhelmingly feminized boys" had mothers with the "same expression of 'bisexuality' [here meaning identification-wise as behaving masculine and feminine at the same time] that prevented their infant sons from separating from their bodies" (366). So it follows, according to Stoller, that the mother's tomboyish 'bisexuality' and her penis envy caused the transsexualism in the boy. Transsexualism is construed as a psychological phenomenon that is caused by the mother's (and the father's) maladjustment to their gender roles as well as the mother's "bisexual gender identity" (here meaning cross-gender identification) (367). In his reasoning the parents are maladjusted to their gender roles and insecure in their gender identities and, therefore, not able to provide a properly heterosexual Oedipal setting for their child. Stoller concludes that keeping these boys from resolving the Oedipal conflict caused their transsexuality. Stoller reasons that "the essential psychodynamic process of separating themselves from their mothers" prevents these boys from being "able to start the development of a male gender identity" (366). Stoller relates this to the Freudian theory of Oedipal development and opposes the assumption that castration anxiety is responsible for the boys 'pathological' development towards femininity. He states that:

> it is hard to believe that this femininity, which is observable in these children long before the classically described phallic phase—in one case by age one—is the result of the kinds of castration fears that are known to be fully developed only years later. Also, it seems very unlikely that the blissful state of closeness that such a mother produces can cause the boy so much fear of his penis being cut off that he pleads to have it cut off, and that he can preserve his sense of maleness by becoming a female.
>
> (1968b: 101)

Accordingly, Stoller engages in the discussion of transsexuality (in male children) in relation to Freud's theory of the phallic phase (the stage of Oedipal resolution) only to later refute it again in order to establish core gender identity as being present even before the phallic phase sets in. As mentioned before, he states that "by the time of the phallic stage, an unalterable sense of gender identity—a core gender identity ('I am a male'; 'I am a female') has already been established in the normal person" (1964: 22). What does not become clear is how exactly castration anxiety features here. Neither is clear how the boy ends up preserving his sense of maleness by pleading to have his penis amputated.

Modifying the Oedipus Complex

Seventeen years later Stoller proceeds from the assumption that in "the rare case, despite biologically normal genitals and proper sex assignment, core gender identity can still be influenced by such subliminal or unconscious communications from mother to infant" (Stoller, 1985: 13). This relationship is expressed by the circumstance that "as an infant, such a boy [feminine i.e., 'primary trans-sexual'] usually has an excessively, blissful, skin-to-skin closeness with his mother" (16). For Stoller the bottom line is that "Oedipal conflict ...[is] ... needed to produce the character structures, such as masculinity and femininity, that maintain the society" (78). He uses his own concept of core gender identity to rely on but modify the Oedipus complex by Freud as follows: The boy's first love is the mother but before the father comes in, Stoller already sees an:

> earlier stage in gender identity development wherein the boy is *merged with mother*. Only after months does she gradually become a clearly separate object. Sensing oneself a part of mother—a primeval and thus profound part of character structure (core gender identity)—lays the groundwork for an infant's sense of femininity. This sets the girl firmly on the path to femininity in adulthood but puts the boy in danger of building into his core gender identity a sense of oneness with mother (a sense of femaleness). Depending on how and at what pace a mother allows her son to separate, this phase of merging with her leaves residual effects that may be expressed as disturbances of masculinity.
>
> (16)

Stoller adds that "femininity also requires that a girl separate from her mother, but not particularly from her mother's femininity" (16) and consequently postulates a psychological "protofemininity." In fact, he does this for "cultures everywhere." He adds that boys and men have the "need for constant vigilance against their unacceptable yearning to return to the merging in the symbiosis" (17). If the father does not support the dissolution of the symbiosis between mother and son and fails to serve as a "rival but also as a model for masculine identification" the "threats to maleness and masculinity are too severe" and "perversion or neurosis intervenes" (14).

Stoller, in his masculinist agenda, moreover, reinstalls the Freudian postulate of "genital primacy" of which, of course, "masculinity and heterosexuality are essential parts" (Stoller, 1985: 15). Stoller reiterates the Freudian concepts of penis envy and the castration complex and uses them to pathologize the trans-sexual person as a 'failed man.' Stoller's conclusion to this reads as follows:

> in brief, this newer view of gender identity holds that femininity in females is not just penis envy or denial or resigned acceptance of castration; a

woman is not just a failed man. Masculinity in males is not simply a natural state that needs only to be defended if it is to grow healthily; rather, it is an achievement.

(18)

If this development is guaranteed, the boy "will prefer to have, not to be, a woman" (17). And it is at this point where his celebration of heterosexual masculinity kicks off, as for Stoller masculinity "consists of struggling not to be seen by oneself or others as having feminine attributes, physical or psychologic. One must maintain one's distance from women or be irreparably infected with femininity" (18). Femininity, here, reads like a disease the boy has to avoid like death. Masculinity in Stoller's eyes is an achievement, which boys have to accomplish before the phallic phase sets in—"normal masculine boys" in this text are given an ultimate status. A boy can only be healthy if he becomes more or less 'hyper'-natural, aligning his male core gender identity with his masculine gender identity maintaining the distance to everything feminine and female. Stoller has replaced the shaky concept of the biological force as an ingredient of the recipe for core gender identity with another concept—that of a feminine mother who lets her son separate from her body. Through this move, masculinity in the reworked Freudian Oedipal trias becomes an achievement.

In contrast to Freud's conviction that the primary identification is masculine, Stoller believes that all children begin with a female core gender identity obtained from the maternal symbiosis. Therefore, core gender identity results non-conflictually through identification with the mother. Failure to interrupt the maternal symbiosis pre-oedipally may in boys result in permanent gender identity 'disorders' such as transsexualism. So-called normal development facilitates the boy's shift to a male core gender identity and to obtain a masculine gender identity. Stoller therefore, basically claims that femininity creeps into the core gender identity of all males—or, as he put it, that there exists a "protofeminine core" in male core gender identity. This "core," somehow, is simply not as stable as the word "core" suggests (Hughes, 1999: 104). Stoller concludes, that his reworking of the Oedipal conflict has rescued femininity from the Freudian assumption that it is pathological, because "the girl is now seen to have an advantage" (17). He however, sums his modification of the Oedipal conflict up by declaring masculinity "an achievement" (18).

However, Stoller will destabilize his own generalization of this theory of protofemininity and the achievement of masculinity. He refers to a body of anthropological literature, which challenges his theories by observing "prolonged gratification of infants in some primitive societies" (Stoller, 1968b: 106). In these reports mothers and sons are "in a happy skin-to-skin contact for many hours of the day and night, and for years" (106). Here, Stoller finds fault with the anthropological accounts because to him "the reports fail to mention that these customs, which sound the same as what is seen in our transsexual boys, cause excessive identification with mother" (106). Stoller, however, fails, I argue, to see the possibility for rethinking his own theories. Yet, he admits that he does not have the data to deal with this exciting complication. Only direct observation and comparison of the subtle way in which white middle-class American mothers versus these primitive mothers handle, breathe upon, coo over, nestle, hover over, ignore, look at, and otherwise impinge upon their infants can make true controls out of otherwise too gross observations.

(106, 107)

Stoller, therefore, admits that his research lacks a control group from other cultures. He states that his "patients have been primarily white, middle-class Americans" and that he, therefore, has plans "to correct this flaw in the future" (xiv). In fact, he will try to do this together with Gilbert Herdt and Julianne Imperato-McGinley, whose collaborations are the focus of chapters 4 and 5 in this volume. However, my contention is that Stoller failed to reconsider his theory; rather, he intensified the problematic assumptions in his work by going cross-cultural. The research Stoller undertakes with Herdt in Papua New Guinea (Stoller & Herdt, 1985, 1990) and Imperato-McGinley in the Dominican Republic (Imperato-McGinley, 1979a) is equally infiltrated with the notion of hegemonic heterosexual masculinity as *the* natural achievement of culture. In the research with Imperato-McGinley, Stoller again re-establishes his postulated biological force as the pre-natal influence of testosterone. In the collaboration with Herdt, however, his influence on the study can be seen in Herdt's theorizing of a hermaphroditic gender identity for the hermaphroditic body.

Gender and Identity—Shaky Concepts?

Stoller was conscious of the problems, which accompany the concept of identity and thus problematizes the concept of gender identity in his publication *Sex and Gender: On the Development of Masculinity and Femininity* from 1968, which will later be crucial for the development of feminist theory (Stoller, 1968b, see chapter 3 in this volume). He states here that:

> as regards the word identity, my treatment of that word will not be more adequate as to the purpose of this work is not to arrive at a comprehensive or even useful definition of the term *identity*, or to enter into the controversies now very much in the forefront of psychoanalytic theorizing as to the differences and similarities, usefulness of distortions of such terms *as ego, self, self-representations, identity, sense of identity* and the like.

(x)

Stoller adds a footnote in which he states that one of his colleagues reviewed the literature about the term identity and has concluded that it "has little use other than as fancy dress in which to disguise vagueness, ambiguity, tautologies, lack of clinical data, and poverty of explanation" (x). Stoller himself therefore, questions the basis of his own work fundamentally. The concept of identity, as well as the terms *ego, self, self-representations, sense of identity* etc. do not receive a definition or even a demarcation. He admits that "it is a working term" (x) yet fails to engage with as cautiously as it should be resulting. The concept of identity remains an empty signifier in his work. However, one thing he seems certain about is that "though it deals with another realm of feeling, thoughts, and behavior than that encompassed by, say, *sexual activity*, the two terms are contiguous and at times inextricably intermingled" (vi). Identity as a concept is in Stoller's terms not to be delineated from other psychoanalytical concepts, but he is convinced that it is adjacent to and blended with sexual activity. Identity is intrinsically bound up with sexual activity in Stoller's perception. The vagueness, which he accepts when it comes to his central working term is also extended to the concept of gender. Stoller continues that "with gender difficult to define and identity still a challenge to theoreticians, we need hardly insist on the holiness of the term gender identity" (vi). Surprisingly, Stoller continues in *Sex and Gender* to work with these undefined terms and bases his complete work on them.

Hermaphroditic Gender Identity I

In his 1968 book on *Sex and Gender,* Stoller published a chapter on "The hermaphroditic Identity of Hermaphrodites." In line with his continuing habit to create new categories and to naturalize identity, he invents a new gender with a different core gender identity. He reports from his clinical data that there are people who "do not feel themselves to be members of either one of only two possible sexes" (34). He tries to work through his three factors (the biological force, the infant–parent relationship and the bodily ego) producing core gender identity and refers to the parents of these people, who show uncertainty as to the "true" sex of their children, which causes the patient also to be uncertain. He argues, that a "sense of body configuration that is in fact different from others will produce a different body ego" (34). Stoller concludes that the child "is in that peculiar position of agreeing with all the world that there are, as it says, only two sexes, while he [sic!] belongs to neither" (34). Stoller cannot resist, even though having cautioned against the concept of gender identity in the introduction of the same volume *Sex and Gender*, to install a separate core gender identity for the hermaphroditic person. He states that "such a person, then, belongs to an entity that has not previously been distinguished from other identity problems." Stoller will now do this and create a "third gender" category for the hermaphroditic person. He proposes a "resulting character structure" for the intersexualized person and "special ways" of managing his_her life, which, as Stoller argues, are "in our society at least, (...) evidence of a different core gender identity, and therefore of a different life perspective" (34).

This third core gender identity Stoller postulates is portrayed as fundamentally different from the male and female core gender identity: it is a hermaphroditic core gender identity. However, the postulation of a different life-perspective for the person with a hermaphroditic core gender identity does not result in the possibility of living this different life perspective with a different body. Quite the contrary, Stoller suggests, that s_he "can wait with relative equanimity for the day when he [sic!] will be fixed so that he can belong; or he does not wait, but bows to his fate of not really belonging to the human race" (34, 5). There-fore, according to Stoller, the hermaphrodite does not belong to the *human race*. He states that the only option for a hermaphroditic person to belong to the human race is to be 'fixed.' The human race as a powerful narrative is called upon to delimit what a hermaphrodite is, or rather what s_he is not. This discursive pro-duction of humanness as *the* expression of binary sex(ual) difference is a con-densation of norms that are repeatedly articulated. The person who is neither male nor female—or both—which interestingly does not feature here as an option, is not considered to be a part of humanity. It is the necessary 'outside' of that what qualifies as human—the single sexed and gendered subject.

The 'practical conclusion' Stoller draws from his account on hermaphroditic gender identity is to argue for treatment, which in the case of the third gender, "cannot help but be successful unless grossly mismanaged" (38). This recommen-dation—echoing Money, Hampson, & Hampson (1955b)—testifies in the frame-work of normalization to the method of erasure. Even though Stoller has postulated a third gender identity resulting from a specific bodily configuration his conclusion is to erase it origin—the bodily configuration and material. He bases his argument upon the inclusion of the intersexualized person into the human race. That which does not belong to the human race, the features of 'ambiguity' and abjected char-acteristics, have to be erased through surgery in order for the subject to belong. However, Stoller has to acknowledge that some cases "appear" to live "comfort-ably in alternating genders." He draws the "practical conclusion" that "one must accurately determine the patient's core gender identity. If it has become firmly established, as it more or less is beyond two to three years of life, then it should not be changed" (38). He echoes Money et al.'s theory of surgical intervention in infancy; yet, with his postulation of a hermaphroditic core gender identity he has to caution against feminizing or masculinizing treatment of the child after this identity has been established. This of course does not contradict Money's treat-ment paradigm, it rather enforces it especially in terms of urgency: the psycho-sexual emergency paradigm that requires operating before the first eighteen months. Because, leaving the child alone, might cause it to develop such a her-maphroditic core identity. As sociologist Georgiann Davis and her colleagues report in 2015, medical experts in intersexualization still perpetuate "the creation of the emergency situation" (12). Stoller himself concludes: "The proper sex must be diagnosed as soon as possible: Only by careful and rapid diagnosis can future emotional problems be avoided" (Stoller, 1968: 29). This is astounding because the "proper sex," as should logically follow is the hermaphroditic sex that should be diagnosed. Yet, this is not so—Stoller's proper sex is either male or female—and somewhat more favourably male than female, as we can surmise.

Reducing the Bodily Ego

The element of gender identity that Stoller normally names as the first one in the list of three is the bodily ego. In 1959 he has not yet developed a clear systematics, yet, in 1964 he describes it as follows: "the anatomy and physiology of the external genital organs, by which is meant the appearance of and the sensations from the external, visible, and palpable genitalia" (220). In 1965 this is further elaborated on as follows:

> Part of the infant's awareness of its existence and separateness form the rest of the world (body ego), like its awareness of being a one-headed, four limbed creature with sensitive openings and exits, comes from this growing awareness of the external genitalia. [...] These sensations confirm the ascription of maleness and femaleness by society.
>
> (199)

And he goes on: "With greater mobility, language, experience, and independence, the child learns not only to consider genitalia very important but that genitalia distinguish two (and only two) classes of humans (and other creatures)" (199). In 1986, Stoller gives two accounts on this element of gender identity. One of them reads as follows: "the infant's growing awareness of its external genitalia" (1968a: 364) and the other as "the anatomy of the external genitalia, which serves as a sign to the physician who delivers the infant and to the parents, the child and the community that the ascription of sex was properly made" (Baker & Stoller, 1986: 1653). And in the same year in his account on "Hermaphroditic Identity of Hermaphrodites" he states that a core gender identity is produced "by the infant–parent relationship, by the child's perception of its external genitalia, and by a biological force" (Stoller, 1968b: 30). He adds that factor one and two "are almost always crucial in determining the ultimate gender identity" (30). As shown above, Stoller bases his entire theory of a third—hermaphroditic gender identity on these two factors. In his *Presentations of Gender* from 1985 Stoller however lists five elements for the development of gender identity, the last one of which is body ego, being defined as "one's sense of the dimension, uses, and significance of the body" yet, he adds:

> even when anatomically defective, so that the appearance of the genitals and their sensations are different from those of intact males or females, the individual develops and unequivocal sense of maleness or femaleness if the sex assignment and rearing are unequivocal.
>
> (14)

With this statement, Stoller contradicts his earlier findings on a hermaphroditic gender identity which was supposed to be based on the combination of the bodily ego and the infant–parent relationship. Now, as we can see, the body vanishes as a differentiated source for multiple—well actually only triple—identifications and becomes reduced to only two possible bodily egos. By granting dominance to sex assignment and rearing, Stoller echoes Money et al. (1955b) even more powerfully than before and feeds their argument through sacrificing the psychoanalytical subversive potential which is inherent in the bodily ego.

The bodily ego is a much more complex aspect of identity formation than Stoller admits.

I agree with Judith Butler that "there is not necessarily one imaginary schema for the bodily ego" (1993: 87) and neither are there just two, or three, as Stoller implies; there is a multiplicity. Butler rejects a single identification that consolidates and reifies the ego and univocally characterizes one's sex, gender, and desire. She argues that masculine and feminine morphologies are arbitrary, including the genitals. Butler argues with Freud and Lacan that "the very contours of the body, the delamination of anatomy, are in part the consequence of an externalized identification" and hereby establishes that the "anatomical is only "given" through its signification," whereby it always already exceeds this signification" (90). Butler dissolves any direct link between anatomy—here the genitals—and a simple consecutive identification with maleness and femaleness, as Stoller wants to suggest. Feminist psychoanalyst Jan Campbell, refers to Freud's definition of the ego as "first and foremost a bodily ego; it is not merely a surface entity, but is itself a projection of a surface" (Freud, 1961 [1925]: 26).[8] She states that "the ego is a mental projection of the surface of the body. The ego is not simply a kind of consciousness based on biological and bodily affects; it also contains unconscious bodily affects of experiences" (2000: 52). These unconscious bodily effects are often disregarded in simplifying theories on the body and the bodily ego. The body and the bodily ego, I argue, could rather be understood as an event, or a process of becoming that is not (solely) based on sex(ualiz)ed_gendered experiences of ones materiality (Budgeon, 2003; Braidotti, 2002). A myriad of experiences influences embodiment, which cannot be reduced to one's status as a boy or a girl, neither to one's sex(ualiz)ed_gendered body configurations. Embodiment can hardly be dichotomously produced because unconscious effects are not dichotomously organized.

Intersex Embodiment and Bodily Ego

In intersexualization the effects of surgical intervention on the body, the bodily ego, embodiment or even the subject in the simplest understanding of the word, are constantly neglected. Feminist psychologist Katrina Roen argues that if we understand the body as an event, then any surgery on the infant's body "will be ever-present" in the adult's later embodiment (2009: 21). Intersex activist and scholar Morgan Holmes analogically argues that genital surgery makes bodies *more* intersex than they started out (2002). And Iain Morland, also intersex activist and scholar notes in this regard that "with intersex births the signifiers 'boy' and 'girl' lose their anatomical referents" because "'boy' and 'girl' act out, rather than refer to, a sex difference" (2001: 537). By the means of surgery in intersexualization, medicine therefore creates sexual dimorphism out of intersex* flesh in order to bring healthy but culturally unacceptable flesh into line with the signifiers of 'boy' and 'girl.' Hereby, the even production of a different bodily ego that is produced by surgery is completely ignored. In fact, as Roen argued, the infant is considered to have not "yet *become* a subject to whom the body is an important marker of selfhood" (2009: 21 [emphasis in original]). Stoller's argument that it is the parents' uncertainty, which produces the different bodily ego opens a completely new field which has also recently been covered by Roen (Doyle & Roen, 2008; Roen, 2009). To her, it is adults who project their fears and ideas about normality onto the child. These fantasies of normality are what the child is supposed to embody. This disregards the immense implications that surgery and medicalization will have on the child's sense of self and embodiment. These interventions are likely to dominate the embodiment and the bodily ego rather than giving the child the normality that the parents project onto it: being a 'normal' boy or girl.

Interpellation as a girl or a boy, according to Judith Butler, is the setting of a boundary and hereby the repeated inculcation of a norm. Holmes states in response to this that "one must be able to say either 'I am male' or 'I am female' not 'I am I'" (2008: 96). She adds as an explanation that "this latter statement does not even make sense, is not even thinkable, without the implication of sex—at least, that is the threat and the fear" (96). This argument is part of the power that delimits and sustains "that which qualifies as the human" (Butler 1993: 8). This delimiting process of producing a subject can be seen in the abjection[9] of those beings who "do not appear properly 'gendered'" (8), as Butler states. And according to Julia Kristeva "it is not a lack of cleanliness or health that causes abjection but what disturbs identity, system, order. What does not respect borders, positions, rules, the in-between, the ambiguous the composite" (1982: 4). Thus, the intersexualized person—the one who is properly in-between, either/or, neither/nor, and both—provokes abjection. With the production of somebody as abject, Butler states, "it is their very humanness that comes into question" (8). These productions of normalizing powers are necessarily and forcefully fostered by a repressive violence that constructs a divide between bodies that matter and other 'abjected' bodies. The abject is therefore a valuable tool to dismantle the processes of intersexualization, because it shows how the process of intersexualization, through positioning 'normal' bodies as dependent on 'other' and 'abjected' bodies or identities, constructs the norm (Butler, 1993; Holmes, 2008). In Stoller's accounts normalizing powers of both erasure and production are at work. Repression is most obvious in the method of erasure, but what about the methods of production? What are the reference points for a possible production of the intersexualized subject?

In fact, in Stoller's research the modes of production of the abject achieve a new meaning. A hermaphroditic body produces a hermaphroditic core gender identity if not erased. The concept of core gender identity derives from Stoller's equation of sex(ual) difference with the boundary between gender identities and his denial of the plurality of developmental positions. The unilinear line of development, which he invokes with the concept of core gender identity, is ultimately referable to the anatomical difference. Hereby, Stoller's postulated psychical protofemininity enables the celebration of masculinity. His theory of core gender identity is unthinkable without the construction of femininity as 'other.' Additionally, the construction of a third gender identity is unthinkable without the construction of a system of two against a system of three. The norm of binary sexual difference is naturalized through the inclusion of a third that does not interrupt the two, yet affirms them as the norm. The naming of intersexuality as an "extraordinary phenomenon" or a "natural experiment" (Stoller, 1964: 220) functions as an irritation and produces an affirmative argument concerning sexual dimorphism (Butler, 1993; Klöppel, 2002). The reference to the 'human race' hereby features as a powerful rhetorical instrument. The abject of the hermaphroditic body, through its positioning as the 'other,' establishes and reaffirms what is human and what is not. The concept of the abject can also be used to describe groups of people who are abjected by society. These groups of people are not objects but they are not subjects either. People on the margins of society are abjected because they do not conform to certain norms, such as hegemonic heterosexuality, masculinity, whiteness, ability, or middle-class attitude. Trangender theorist Riki Wilchins explains that the margins of society "are margins because that's where the discourse begins to fray, where whatever paradigm we're in starts to lose its explanatory power and all those inconvenient exception begin to cause problems" (2004: 71). Wilchins, furthermore, explicates that at these margins "science no longer asks but tells. Nature no longer speaks the truth, but is spoken to" (79).

In Stoller's theories, this mechanism becomes clear; the norm is reinstalled by reference to the 'natural abnormality' of the hermaphrodite. The relation between the natural and the normal is in itself contradicting. In short, the 'normal body' is diametrically opposed to the 'natural body,' which is paradoxical in that Western discourses are organized as conceiving of the normal and the natural as one and the same. Thus, 'other' bodies and identities become invisible through the techniques of normalizing bodies which are at the same time made to appear natural (see for example Holliday & Hassard, 2001). The intersexualized body with the accompanying hermaphroditic core gender identity is one of them. If a male body produces masculinity and a female body produces femininity then a problem arises if a hermaphroditic body is acknowledged. Because the two other core gender identities are said to be produced by the body and are referred back to the body, then a hermaphroditic body disrupts this logic. Subsequently, the hermaphroditic body has to produce a hermaphroditic core gender identity in order to not threaten but reaffirm the naturalization of core gender identity as dimorphic. The postulation of a hermaphroditic gender identity for a hermaphroditic body does not put the binary framework into question; it reaffirms it.

Conclusion

At the beginning of his lifelong study of 'intersex patients,' Stoller worked with the concepts of 'sexual identification' and 'sexual identity.' For him, this meant identification with one sex or the other—"a clearcut commitment to one sex" (Stoller & Rosen, 1959: 265). A commitment to one sex is in my eyes something very different than an 'unalterable sense of being male or female' as the concept of 'core gender identity' declares. The first one implying a warranted decision that could be changed willingly by someone, the second suggesting an unalterable core, that inhabits and dictates the person.

After some time and several case studies that talked rather cautiously about processes of identification, Stoller shifted from 'sexual identity' to 'core gender identity.' From his point of view, the disadvantages of the first concept were that 'sexual identity' implied sexual activities or fantasies. The advantages of switching to 'core gender identity' were that he could now frame 'core gender identity' in terms of a so-called biological force. This new 'core gender identity' allowed him to attribute the development of the supposed 'core' to a "silent component: a congenital, perhaps inherited biological force" (1964: 225). On the one hand, this core seems not as stable as the word suggests, because identification with the 'opposite sex' can always creep in, on the other hand, the 'core' is used to argue for early surgery. Even though Stoller started out by cautioning against hasty decision making *for* the intersex patient he will more and more in the course of his career tend towards very early sex assignment as male or female.

However, even though Stoller framed the bodily ego as one of the three elements in gender identity, the body as the source of pleasure and meaning, was intriguingly absent from his theoretical discussion of gender (see also Hughes, 1999: 158). He tied sexual object-choice to gender, with heterosexuality as the normal accompaniment to a core gender identity that at least in girls was supposedly conflict-free. This allowed him to proclaim some kind of protofemininity present in all children and masculinity as an achievement. Here, transsexual children featured in his research, which he used to support his theory of pathological mother-son-relationships. Surprisingly even though he argued with the Freudian concepts of drive, and the Oediapl trias, erotic desire was never—apart from the pathologization of homosexuality—mainly via the butch—considered as something fundamental.

His interest in intersexualizing children brought Stoller to transsexualizing children. In making the mother responsible for a failed masculinity and referring to her as tomboy or bisexual, Stoller could reinstall the heteronormative Oedipus complex, yet modify it considerably. He reconfigured the triangularity of the Oedipus complex into a basic structure of protofemininity. This protofemininity has to be overcome by the struggling masculine component at the cost of a loving mother-son relationship. What is made impossible is a masculinity which can cope with intense bodily contact (with the mother). He not only reinstalls the Oedipus complex as the precondition for patriarchal culture but also prevents any other narrative that could be inclusive enough to accommodate more than dichotomously organized genders. The intersexualized body threatens the concept of a dimorphous biological force that exclusively produces neatly and binary gendered core identities. In order to remove this threat, Stoller postulated a hermaphroditic identity on the basis of a hermaphroditic body. What can be named and abjected can also be easily kept at bay. Stoller established *a hierarchy of normality in nature* in intersexualization.

Notes

1. Harold Garfinkel's work was important for feminist scholars. Suzanne Kessler and Wendy McKenna described Garfinkel's contribution as a methodological account of the nature of the social production of gender (Kessler & McKenna, 1978) and as having provided an example of the ethnomethodological method that can reveal how stable, accountable, practical activities are produced by people in everyday life. However, critique has also been aired. Norman Denzin argued from a critical feminist perspective that Garfinkel's assessment of Agnes (which Stoller also worked on as is discussed below):

   > has much in common with conventional interpretive sociology, in which there is a masculine preoccupation with theorizing the genesis, origins, causes, and effects of various social situations, including social problems and the types of persons and groups who have or who are those problems.
   >
   > (1990: 198)

2. German original:

   > Unter einem "Trieb" können wir zunächst nichts anderes verstehen als die psychische Repräsentanz einer kontinuierlich fließenden, innersomatischen Reizquelle, zum Unterschied vom "Reiz," der durch vereinzelte und von außen kommende Erregungen hergestellt wird. Trieb ist so einer der Begriffe der Abgrenzung des Seelischen vom Körperlichen. [...] Die Quelle des Triebes ist ein erregenden Vorgang in einem Organ, und das nächste Ziel des Triebes liegt in der Aufhebung dieses Organreizes.
   >
   > (Freud, 2000 [1905]: 77)

3. The "Agnes case" is described in Harold Garfinkel's ground-breaking text, *Studies in Ethnomethodology* (1967) in full length. Garfinkel worked on this case with Stoller at the Department of Psychiatry of the University of California, Los Angeles (UCLA). Garfinkel's account of Agnes has been accorded the status of a sociological "classic" (Denzin, 1990, 1991).

4. Since then there have been numerous attempts to question the assumed universality of the Oedipus complex. Judith Butler has shown with Antigone, Oedipus's sister: if Freud had chosen another myth to identify the workings of our culture then we would have a different psyche (Butler, 2000). Her theory is that the gods cursed Oedipus and his children, and psychoanalysis curses gender (Butler, 2000). Other theorists have engaged from a different angle with Freud's universal theory, stating that it is the work of mourning (his father's death) and that if Freud had stuck with his earlier findings he would have had to consider not only the Oedipus part of the myth but also other parts such as Laius's (Oedipus's father's) fault (Balmary, 1979) or his mother and wife Iocasta (Silverman, 1988). Also, he has focused on exclusively one version of the myth and not the others, which is problematic because according to Claude Lévi-Strauss "all versions belong to the myth and should be considered" (Lévi-Strauss, 1963: 206–231).

5. Elektra is also a mythical figure. She was the daughter of Agamemnon and Clytemnestra and helped her brother Orestes to kill her mother.

6. Freud's contribution to the understanding of patriarchal culture is by now acknowledged by most feminist scholars, and his theories are sometimes even regarded as subversive, because he described what patriarchy demands of women and femininity and thereby does not argue with biology (e.g., Mitchell, 1974). Among some, psychoanalysis is, therefore, regarded as a powerful tool to dismantle ideological/social differences between the sexes/genders. However, a number of feminists have argued that Freud's emphasis on white male development and his disregard of or inability to sufficiently describe female sexuality and development affirms patriarchal structures and therefore phallocentrism (e.g., Irigaray, 1981).

7. In fact, as Brian Loftus argues, "bisexuality works as a semiotic tool to naturalize heterosexuality" (Loftus, 1996).
8. German original: "Das Ich ist vor allem ein körperliches, es ist nicht nur ein Oberflächenwesen, sondern selbst die Projektion einer Oberfläche" (Freud, 2000 [1923]: 294).
9. Julia Kristeva introduced the concept of the abject in psychoanalysis (Kristeva, 1982). Abjection is derived from the Latin term *abjicere*. It means to expel, to cast out or away. According to Kristeva, the abject threatens the superego, which means that the abject is a threat to the ordering law of the subject or the self. A social being is constituted through the force of exclusion of impure entities such as excrement, menstrual blood, urine, semen, tears vomit, food, masturbation, or incest. The self has to erase these elements in order to become a social self. For Kristeva, the process of abjection, however, can never be fully completed because the excrement, menstrual blood, urine, semen, tears vomit, food, masturbation, and incestuous entities can never be fully obliterated. The self is haunted by these elements; it is constantly threatened to be disrupted or even dissolved. Abjection marks the borders of the self, yet it also always threatens the self constantly. Freud was the first to suggest that civilization is founded on the repudiation of certain pre-oedipal pleasures i.e., the polymorphously perverse and incestuous attachments (Freud, 1927).

Bibliography

Angelides, S. (2001). *A history of bisexuality*. Chicago, London: The University of Chicago Press.

Baker, H. J., & Stoller, R. (1968). Can a biological force contribute to gender identity? *American Journal of Psychiatry, 124*(12), 1653–1658.

Balmary, M. (1979). *Psychoanalyzing psychoanalysis. Freud and the hidden fault of the father*. Baltimore, London: The Johns Hopkins University Press.

Braidotti, R. (2002). *Metamorphoses: towards a materialist theory of becoming*. Cambridge: Polity Press.

Budgeon, S. (2003). Identity as an embodied event. *Body and Society, 9*(1), 35–55.

Butler, J. (1990). *Gender Trouble. Feminism and the Subversion of Identity*. New York: Routledge.

Butler, J. (1993). *Bodies that matter: on the discursive limits of 'sex.'* New York: Routledge.

Butler, J. (2000). *Antigone's claim, kinship between life and death*. New York: Columbia University Press.

Campbell, J. (2000). *Arguing with the Phallus: Feminist, queer and postcolonial theory: A psychoanalytic contribution*. London: Zed Books.

Denzin N. K. (1990). Harold and Agnes: A feminist narrative undoing. *Sociological Theory, 8*(2), 198–216.

Denzin, N. K. (1991). Back to Harold and Agnes. *Sociological Theory, 9*(2), 280–285.

Doyle, J. and Roen, K. (2008). Surgery and embodiment: Carving out subjects. *Body and Society, 14*(1), 1–7.

Freud, S. (1961 [1905]). Three essays on the theory of sexuality. In *The standard edition of the complete psychological works of Sigmund Freud* (Vol. 7), (James Strachey, Trans. and Ed.). London: Hogarth.

Freud, S. (2000 [1905]). *Drei Abhandlungen zur Sexualtheorie*. Frankfurt am Main: Fischer Taschenbuch Verlag.

Freud, S. (2000 [1923]). *Das Ich und das Es.* Frankfurt am Main: Fischer Taschenbuch Verlag (Studienausgabe, Band III).

Freud, S. (1953 [1913]). Totem and taboo. In *The standard edition of the complete psychological works of Sigmund Freud* (Vol. 13), (pp. ix-163), (James Strachey, Trans. and Ed.). London: Hogarth.

Freud, S. (1961 [1925]). The Ego and the id. In *The standard edition of the complete psychological works of Sigmund Freud* (Vol. 19), (pp. 1–66), (James Strachey, Trans. and Ed.). London: Hogarth.

Freud, S. (1961 [1927]). Civilization and its discontents. In *The standard edition of the complete psychological works of Sigmund Freud* (Vol. 21), (James Strachey, Trans. and Ed.). London: Hogarth.

Freud, S. (1961 [1930]). Civilization and its discontents. In *The standard edition of the complete psychological works of Sigmund Freud* (Vol. 21), (pp. 59–147), (James Strachey, Trans. and Ed.). London: Hogarth.

Freud, S. (1937). Analysis terminable and interminable. *International Journal of Psycho-Analysis, 18*(4), 373–405.

Garfinkel, H. (1967). *Studies in ethnomethodology.* Englewood Cliffs: Prentice Hall.

Hampson, J. (1955). Hermaphroditic genital appearance, rearing and eroticism in hyperadrenocorticism. *Bulletin of Johns Hopkins Hospital, 96*(6), 265–273.

Herdt, G, & Stoller, R. (1985). Sakulambei—A hermaphrodite's secret: An example of clinical ethnography. *Psychoanalytic Study of Society, 11,* 117–158.

Herdt, G., & Stoller, R. (1990). *Intimate communications: Erotics and the study of culture.* New York: Columbia University Press.

Holliday, R., & Hassard, J. (Eds.). (2001). *Contested bodies.* New York: Routledge.

Holmes, M. (2002). Rethinking the meaning and management of intersexuality. *Sexualities, 5*(2), 159–180.

Holmes, M. (2008). *Intersex: A perilous difference.* Selinsgrove, Pennsylvania: Susquehanna University Press.

Hughes, J. (1999). *Freudian analysts/feminist issues.* New Haven: Yale University Press.

Imperato-McGinley, J., Peterson, R. E., Gautier, T., & Sturla, E. (1979). Androgens and the evolution of male-gender identity among male pseudohermaphrodites with 5-alpha reductase deficiency. *New England Journal of Medicine, 300*(22), 1233–1237.

Irigaray, L. (1981). *This sex Which is not one.* Ithaca, New York: Cornell University Press.

Kessler, S., & McKenna, W. (1978). *Gender: An ethnomethodological approach.* London: The University of Chicago Press.

Kiefer, J. H. (1957). Recent advances in the management of the intersex patient. *The Journal of Urology, 77*(3), 528–536.

Klöppel, U. (2002). "Störfall" Hermaphroditismus und Trans-Formationen der Kategorie "Geschlecht." Überlegungen zur Analyse der medizinischen Diskussionen über Hermaphroditismus um 1900 mit Deleuze, Guattari und Foucault. *Potsdamer Studien zur Frauen- und Geschlechterforschung, 6*(2), 137–150.

Kristeva, J. (1982). *Powers of horror. An essay on abjection.* New York: Columbia University Press.

Levi-Strauss, C. (1963). *Structural anthropology.* New York: Basic Books.

Loftus, B. (1996). Biopia: Bisexuality and the crisis of visibility in a queer symbolic. In D. Hall and M. Pramaggiore (Eds.). *RePresenting bisexualities: Subjects and cultures of fluid desire* (pp. 207–233). New York: New York University Press.

Money, J. (1955). Hermaphroditism, Gender and Precocity in Hyperadrenocorticism: Psychologic Findings. In *Bulletin of Johns Hopkins Hospital, 96*(6), 253–264.

Money, J., Hampson, J. G., & Hampson, J. L. (1955a). An examination of some basic sexual concepts: The evidence of human hermaphroditism. *Bulletin of Johns Hopkins Hospital, 97*(4), 301–319.

Money, J., Hampson, J. G., & Hampson, J. L. (1955b) Hermaphroditism: Recommendations concerning assignment of sex, change of sex, and psychologic management. *Bulletin of Johns Hopkins Hospital, 97*(4), 284–300.

Money, J., Hampson, J. G., & Hampson, J. L. (1956). Sexual incongruities and psychopathology: The evidence of human hermaphroditism. *Bulletin of Johns Hopkins Hospital, 98*(1), 43–59.

Money, J., Hampson, J. G., & Hampson, J. L. (1957). Imprinting and the establishment of gender role. *A.M.A. Archives of Neurology and Psychiatry, 77*(3), 333–336.

Morland, I. (2001). Is intersexuality real? *Textual Practice, 15*(3), 527–547.

Oakley, A. (1972). *Sex, gender and society.* London: Temple Smith.

Roen, K. (2009). Clinical intervention and embodied subjectivity: Atypically sexed children and their parents. In M. Holmes (Ed.), *Critical intersex* (pp. 15–40). Farnham: Ashgate.

Silverman, K. (1988). *The acoustic mirror. The female voice in psychoanalysis and cinema.* Bloomington, Indianapolis: Indiana University Press.

Stoller, R. (1964). A contribution to the study of gender identity. *International Journal of Psychoanalysis, 45*(2–3), 220–226.

Stoller, R. (1965). Passing and the continuum of gender identity. In J. Marmor (Ed.), *Sexual inversion: The multiple roots of homosexuality* (pp. 190–210). New York: Basic Books.

Stoller, R. (1968a). A further contribution to the study of gender identity. *International Journal of Psychoanalysis, 49*(1), 364–367.

Stoller, R. (1968b). *Sex and gender. On the development of masculinity and femininity.* New York: Science House.

Stoller, R. (1985). *Presentations of gender.* New Haven, London: Yale University Press.

Stoller, R., & Rosen, A. C. (1959). The intersexed patient. *California Medicine, 91*(5), 261–265.

Stoller, R., Rosen, A. C., & Garfinkel, H. (1962). The psychiatric management of intersex patients. *California Medicine, 96*(1), 30–34.

Wilchins, R. (2004). *Queer theory, gender theory. An instant primer.* Los Angeles: Alyson Publications.

3   From Five Sexes to n-sexes

In the process of intersexualization, John Money's gender role and Robert Stoller's gender identity (see Chapters 1 and 2 in this volume) had wide-reaching consequences for feminist thinking. In fact, much of feminist theory is indebted to Money's and Stoller's coinage of the gender-concept; it was first introduced to feminist thought by Ann Oakley's widely read *Sex, Gender and Society* (1972). Gender, as something distinct from sex has since enabled feminist theory to shed light on sex(ual) difference as historically variable, and consequently as dependent on gendered conceptualizations in the bio-medical disciplines. For more than 40 years, feminist scholars have continued to develop gender as an analytical tool to interrogate social, political, and cultural formations. Feminist theorists from all over the world have happily engaged in using the gender-concept as a tool to dismantle patriarchy, (hetero-)sexism, and discrimination. The gender-concept, however, based on the distinction between sex and gender helped establish the pathologization of individuals who express incoherence between the two. This pathologization results in intersexualization and has one unquestionable backdrop—even though intersex questions it fundamentally—the naturalness and normality of dimorphic sex. The problems that arise from this fact are not just to be seen in feminist theorization but also of course in the treatment paradigms of intersexuality that are still in place (see Chapter 1 in this volume).

More than 20 years ago, Moira Gatens, feminist from Australia, considered Robert Stoller to be "the authoritative source for the recent prominence of writings centering on gender" (1991: 141). Toril Moi, feminist native in Norway, stated that Stoller "fired feminists' imagination" when he "medicalized" sex and made "gender" a "purely psychological category" (1999: 22). It was (and still is) considered very helpful to study discursive gender apart from material sex, because sex was less easy and convincing to deconstruct if one did not come from inside the bio-medical disciplines. For some years now there has also been considerable effort from inside history, philosophy, and other humanities to tackle sex as a phenomenon that should not be left solely to the natural sciences. More recently feminist materialism has engaged with the distinction of sex and gender in order to surpass the pitfalls, which come with the gender-concept.

As regards intersexualization, my main aim in this chapter is to show how intersexualization has first helped establish the distinction between sex and gender and second helped develop gender as an analytical tool for many feminist endeavors. Third, the remainder: sex, however has also been interrogated and I aim at somewhat cursorily showing how feminist engagement with sex and the body looks like when coming from either the humanities, here namely Judith Butler's approach or the biological sciences. The historicization of the body coming from a variety of disciplines has done important conceptual work, which has also helped deconstruct sex and subsequently also intersexualization. The entanglements between the gender-concept, the history of the body, and inter-sexualization are hereby of special interest.

As regards the connection between intersexualization, the gender-concept, and feminist scholarship a review of the influential works of feminist biologist Anne Fausto-Sterling follows. Of specific interest is here her proposal that there is biological basis for identifying not two but five sexes. Fausto-Sterling in fact, in large part initiated the intersex movement. Her work was groundbreaking in many ways; whereas she remained dependent upon the dimorphic sex_gender distinction and was thus unable to escape its predicaments in 1993, she ulti-mately escaped binary conceptualizations of sex and gender in 2000. After showing the limitations of her work in 1993, I turn to new ideas she champions in 2000 as well as to theories developed by material feminists. Those theories point to promising ways upon which we might eventually overcome the distinc-tion between sex and gender and therefore also intersexualization—at least in a conceptual way. My claim is that overcoming the binary distinction between sex and gender is ultimately necessary if we are to intervene into intersexualization in the future. Anne Fausto-Sterling's proposal of the five sexes will hereby serve as an emblematic approach for the biological disciplines. Fourth, towards the end of this chapter I will engage with new feminist materialists approaches because they seem promising as regards overcoming the implicit binaries inher-ent in the gender-concept. What has surely motivated this chapter is my own indebtedness to the gender-concept as a feminist scholar of the beginning of the twenty-first century.

The Gender-Concept: Origins, Heritages, and Deconstructions

Insights gained from gender analyses—particularly when in conversation with race, ethnicity, nationality, age, ablebodiedness, and class—have irreplaceable value. However, many feminist and queer theorists overlook the genealogy of the gender-concept, which from its beginning *always* implied dimorphic sex. Gender was and is unthinkable without sex from its very first instantiation by Stoller and Money. Ann Oakley put it in 1972 as follows:

> Whereas Stoller talks about "gender identity," Money and the Hampsons refer to "psychosexual orientation": the meaning of both terms is the sense an individual has of himself or herself as male or female, of belonging to one or the other group. The development of this sense is essentially the same for both biologically normal and abnormal individuals, but the study of the

biologically abnormal can tell us a great deal about the relative parts played by biology and social rearing; there are a multitude of ways in which it can illuminate the debate about the origin of sex differences.

(Oakley, 2015 [1972]: 116)

Oakley therefore champions the gender-concept as particularly helpful in the relationship between biology and the social. Oakley however, in her enthusiasm for the gender-concept reifies the 'normality' of dimorphic sex, which she considers to be "the origin of sex difference"—her argument is that the sense of belonging to the male or female group will also be developed in the 'abnormal.' Thereby installing not just sex but also gender as binary. The nature/nurture debate, which still circles around the relationship between the two, has sprung especially from John Money's work as polemically discussed in John Colapinto's account on David Reimer's case (2000). This case, also known as the John/Joan case has had wide-reaching consequences in the polarization of feminist theory. The whole debate supported essentialist argumentations that are still in place. Intersexualization was *the* single process, which helped establish the gender-concept in feminist thought and was, as we can see in this quote not just relied upon but also fostered through the applied argumentation for the concept.

Jan Campbell's book *Arguing with the Phallus* is a complex account of psychoanalysis, queer theory, and postcolonial theory. Campbell discusses the influence of Stoller's distinction between sex and gender on feminism and states that his:

> work was claimed as a breakthrough by those who wanted to advocate the social construction of sexuality. By arguing that an individual's gender identity was basically a product of post-natal psychological influences, he made biological arguments for sexual identity redundant. Stoller's work was not simply taken up by psychoanalytical feminists such as Chodorow and Dinnerstein, but also by radical feminists, such as Kate Millet and social feminists such as Michele Barrett.
>
> (2000: 26)

Whereas Campbell is correct in her estimation of Stoller's influence, her claim that Stoller never made biological arguments for sexuality redundant is not quite accurate. In fact, Stoller believed in a biological force that influenced gender identity, and he was convinced that our core gender identity is exclusively male or female. Later, however Stoller did develop a third possibility of hermaphroditic gender identity, which nonetheless had to be prevented by surgery in early infancy, thereby echoing Moneys rigid treatment recommendations. For Stoller and Money, gender identity and role was and should be based upon and be congruent with the binary sexed body.[1] Whereas they acknowledged that environmental factors such as upbringing could influence one's gender identity development, they ultimately believed that gender identity is a natural result of biological sex, and that in humans, sex is dimorphic. That is, male and female are two distinct forms that are differentiated by gonads, internal and external genitals, chromosomes, and genes as well as secondary sex characteristics such as breasts, muscle mass, height, hair distribution, larynx, and so on. Gender, for Stoller and Money, follows the so-called natural dimorphic fact of sex, and infants seen to be male at birth should grow into men (read: heterosexual men), whereas infants seen to be female at birth should grow into women (read: heterosexual women). All this, despite their samples of individuals proved exactly the opposite: sex is neither easy to distinguish, nor is it binary. Moreover, in the familiar words of Judith Butler, sex is always and already gender:

> The appearance of an abiding substance or gendered self, what the psychiatrist Robert Stoller refers to as "gender core," is thus produced by the regulation of attributes along culturally established lines of coherence. As a result the exposure of this fictive production is conditioned by the deregulated play of attributes that resist assimilation into the ready made framework of primary nouns and subordinate adjectives.
>
> (1990: 32, 33)

Butler refers to *Presentations of Gender* (1985) by Stoller and develops from this her famous claim that "gender is performatively produced—yet compelled by the regulatory practices of gender coherence" (1990: 33). "Gender," her prominent statement reveals is "always a doing" and she continues "there is no gender identity behind the expression of gender." In her view "identity is performatively constituted by the very 'expressions' that are said to be its results" (33). Thereby refuting completely Stoller's claim of a gender identity that might be there before its expression. Instead, Butler argues that gender is *performative*, meaning that the sexes have no intrinsic, ontological validity, and that gender is not a natural part of a person but something that has to constantly be enacted according to the norms.

Both Butler and Foucault agree that sexuality is the primary political field that necessitates both the normalization of sex and the gendered social order. Butler identifies this as the *heterosexual matrix*—"that grid of cultural intelligibility through which bodies, genders and desires are naturalized" (1990: 151). She states that "the notion that there might be a 'truth' of sex, as Foucault ironically terms it, is produced precisely through the regulatory practices that generate coherent identities thorough the matrix of coherent gender norms" (17). Transposing the Nietzschean idea that there is no "doer behind the deed," Butler argues that there is no gender identity behind the expressions of gender; identity is constituted by and through the very expressions that are said to be its results. Or in Butler's own words, "gender is a matter of doing and its effects rather than an inherent attribute, an intrinsic feature" (17). Butler's reconceptualization of gender as "performative reiteration" opened up new possibilities that had been foreclosed by the uses of the gender-concept and its inherent sex_gender distinction. Butler's theories of *performative subjectivity* provide a powerful set of explanations of the ways in which subject(ivitie)s are constituted through discourses and cultural technologies like bio-medicine. *Gender Trouble* effectively places the neutrality and naturalness of the biological body in question and offers an entirely new theoretical framework for thinking about gender and identity, namely as ritualized repetition. Butler's claim to think of sex only in terms of gender has definitely fostered the deconstructive enterprise.

Butler herself however, realized that she might be misunderstood in this claim and has distanced herself from a purely constructivist view in her subsequent volume *Bodies That Matter* (1993). Butler here tackles the challenge to approach sex from a radical constructivist position. She describes this jeopardy as follows:

> When the sex/gender distinction is joined with a notion of radical linguistic constructivism, the problem becomes even worse [compared to gender as a term that absorbs "sex"] for the "sex" which is referred to as prior to gender will itself be a postulation, a construction, offered within language, as that which is prior to language, prior to construction. If gender is the social construction of sex, and if there is no access to this "sex" except by means of its construction, then it appears not only that sex is absorbed by gender, but that "sex" becomes something like a fiction, perhaps a fantasy, retroactively installed at a prelinguistic site to which there is no direct access.
>
> (5)

Therefore, it is the matter of sex, which needs to be addressed. The questions she tackles in *Bodies that Matter* are therefore concerned with this matter and the normalizing forces with produce it:

> To claim that sex is already gendered, already constructed, is not yet to explain in which way the "materiality" of sex is forcibly produced. What are the constraints by which bodies are materialized as "sexed," and how are we to understand the "matter" of sex, and of bodies more generally, as the repeated and violent circumscription of cultural intelligibility. Which bodies come to matter—and why?
>
> (1993: xi, xii)

For Butler, the biological body is a social body from the very beginning. The body is only accessible through the layers of discourse that surround it. It is the norm that becomes materialized though the citation of normative and symbolic laws. Because it is subjected to the symbolic order, the body materializes as a social entity. The body is the effect of norms; moreover, the symbolic order and society in general materialize in and through the body. Through the performative reiteration of normative bodies, their shapes, their pictures, their perception *and* their morphologies are produced. Materiality is therefore nothing else than the materialization of norms. In Butlerian parlance, the body is the product of power and its technologies and practices. The body is not natural and, therefore, no body can be unnatural. Nothing about bodies can be named natural, not even its inevitable mortality (even this is arguable since the development of technological practises such as cryogenics and gene therapies).

Even though Butler fundamentally deconstructed the category of gender, she still uses it and builds her arguments—though very successfully—around it. Sex, or rather sexual difference as a material and biological entity however, is not accounted for in Butler's philosophy. Others however, do this: feminist biological scholarship has tackled the notion of sex in biological knowledge production already from a number of angles. The body has been historicized, showing that there is no ahistorical body that transcends historical and cultural specificity. More specifically, the sexed body has been examined such that we can no longer claim sex to be the biological fact upon which gender is based. It has been shown, that the body has a history, and the sex(ualiz)ed_gendered body is intrinsically bound up with the social and political circumstances in which it is dissected, interrogated, examined, diagnosed, and treated. This insight is important in many conversations, but nowhere more important than when it comes to intersexualization.

Historicization of the (Sexed) Body

The dawn of theories about the body and its construction can be located at the beginning of the twentieth century. Key figures of the historicization of the body provided us with very different conceptualizations and contextualizations. For example, Maurice Merleau-Ponty's approach in the 1940s was phenomenological and anthropological. He was the first to talk about the conception of the 'lived body.' His emphasis was on the investigation of the body as an active participant in the creation of the self. Merleau-Ponty gave to the body an ontological status in respect to the location of subjectivity in the body and not in the mind or consciousness (1962). French sociologist Marcel Mauss developed the concept of *body techniques* and made possible a different approach to the body as the result of a whole range of material practices, objects, and skills, devised and used by humans in interaction with each other and their surroundings. Mauss interrogated the relationship between symbolic systems, social structures and the body, while analyzing the way cultural systems relied on the body's expressive resources to formulate social relations. In his terms, the human body was always defined according to cultural beliefs about social relations (1936). German sociologist Norbert Elias developed the concept of civilization in which he described the body as an object of disciplining mechanisms (1978, 1982). Last but not least, Michel Foucault greatly influenced the theorization of the body and has, in fact, become inescapable for historians of the body. Foucault shifted the attention from language to discourse. He regarded discourse as a system of representation, of language *and* practice. In his later work, he focused on the relationship between knowledge and power and placed the body at the center of the struggles between different formations of power and knowledge (e.g., 1965, 1978). He introduced the term of *discipline* in respect to the formation of the individual; his work was groundbreaking in respect to discursive formations of the body. His understanding of normalization in which the body is central, is described in *Discipline and Punish: The Birth of the Prison* (1977). Here,

Foucault concludes that "the power of normalization enforces homogeneity; but it individualizes by making it possible to measure deviations, to set levels, to define specialties, and to render differences useful by calibrating them one to another" (182–184). The Foucauldian notion of power is that it is "proportional to its ability to hide its own mechanisms" (1978: 86). Yet, disciplinary power is also constitutive of the body.

Especially, recent research into the history of medicine is often based on Foucault, and has contributed particularly to a new understanding of the discursive formation of the body through medical discourse (e.g., Levin & Salomon, 1990). Using a Foucauldian analysis, scholars started approaching gendered_ sexed and medicalized bodies through the context of language and practices (e.g., Stoff, 2001; Lorenz, 2000; Feher, 1989). This also enables a different approach to the intersexualization of some bodies. Foucault placed the body at the center of struggles between formations of power and knowledge. He describes the following:

> Biological theories of sexuality, juridical conceptions of the individual, forms of administrative control in modern nations, led little by little to rejecting the idea of a mixture or the two sexes in a single body, and consequently to limiting the free choice of the indeterminate individual. Henceforth, everybody was to have one and only one sex. Everybody was to have his or her primary, profound, determined and determining sexual identity; as for the elements of the other sex that might appear; they could only be accidental, superficial, or even quite simply illusory.
>
> (1980: viii)

To Foucault, the body is a bio-political truth and medicine is a bio-political strategy that infiltrates the body at several levels and produces it as that what we understand as the human sexed_gendered body. A body that has to have one sex and only one is easy to identify, to categorize and therefore to control, as Butler describes it:

> Foucault engages in a reverse-discourse which treats "sex" as an effect rather than an origin, In the place of "sex" as the original and continuous cause and signification of bodily pleasures, he proposes "sexuality" as an open and complex historical system of discourse and power that produces the misnomer of "sex" as part of a strategy to conceal and, hence, to perpetuate power relations.
>
> (1990: 121)

At around the same time Foucault was developing his theory of bio-power, thinkers from the field of Science and Technology Studies were using different methodologies and disciplinary knowledge to show that scientific 'facts' do not exist outside of language, history, politics, and culture. Scientific entities and facts such as genes and chromosomes are produced and interpreted by social beings in culturally, geo-politically, and historically situated formations. (e.g., Latour & Woolgar, 1979). All of these bodily entities or substances are of course of great importance to the sexing and gendering of bodies–to the definitions of binary sex(ual) difference.

The historicization of the body with regard to sex(ual) difference has, for instance, also been put forward by Thomas Laqueur, who traces the process of anatomical dichotomization, which he calls the shift from the *one-sex model* to the *two-sex model* (1990). Laqueur argues that this change in perception was motivated by a political agenda and not by new medical insights into anatomy. He shows that sex(ual) difference is always a consequence of the social, political, and cultural construction of gender determining what counts anatomically as sex. He states that:

> during much of the 17th century, to be a man or a woman was to hold a social rank, to assume a cultural role, and not to be organically one or the other of two sexes. Sex was still a sociological, not an ontological, category.
>
> (142)

Laqueur debates the significance of different approaches to the body in different historical periods, including early Christianity, the Early Modern Period, and the nineteenth century. However, Laqueur's distinction between gender and sex might not be appropriate for the Early Modern Period because this constructs a notion not embraced during that time.[2] According to Laqueur, it was during these periods that the shift from the one-sex model to the two-sex model took place. Gender dichotomy, therefore, in Laqueur's view, is not something inevitable, dictated by the natural fact of the two sexes. Concerning the investigation of the perception of hermaphroditism, Laqueur concludes that "maleness and femaleness did not reside in anything particular" (135). "Thus," he reasons, "for the hermaphrodites the question was not 'what sex they are really' but to which gender the architecture of their bodies most readily lent itself" (135). His groundbreaking study has encouraged the notion of the constructed nature of the bodies as sexually dimorphic. His account has powerfully shown that biological explanations of the body are deeply rooted in the political, cultural, and social orders of the world.

The Body in Feminist Theory

Since the 1980s, feminists have challenged biological knowledge production about women's bodies, and they have critiqued the common sense assumption that sex is a biological fact that the social category of gender is based upon. Feminist philosopher, Jane Flax for example, has argued that "biology and nature are rooted in social relations; they do not merely reflect the given structure of reality itself [...] feminist theorists need to deconstruct further the meanings we attach to biology/sex/gender/nature" (1990: 50). Feminist biologists have taken up this challenge and refer to several strands of theoretical thinking.

Critical investigation into the construction of bodies as well as sex(ual) differ-ence has a long tradition and is associated with influential authors (e.g., Bordo, 1993; Birke, 1999; Fausto-Sterling, 1985; Haraway, 1991). Feminist body history means the historicizing of the production and perception of not just bodies of women but also those of men or rather how the two become distin-guished by biological knowledge productions. What becomes obvious in these accounts is that the body is a multidisciplinary subject and that the history of its discursive construction only becomes intelligible through the analysis of several distinct mechanisms and technologies. Because I understand intersexualization as the quest for a scientifically verifiable distinction between the sexes, an unfolding of feminist engagements in the deconstruction of these scientific modes seems now to be in place. In order to pay justice to the work that the gender-concept was however able to do for feminist endeavors some prominent examples will be considered in the following paragraphs.

Science and technologies scholar Nelly Oudshoorn documents how hormones were first conceived of as sexually specific. Intersexualization, in fact, led to hormone research that produced testosterone as male and oestrogen as female. It was an endocrinologist who coined the term intersexuality in 1916 as shown in the introduction (Goldschmidt, 1916). Carol Worthman argued that hormonal action is mediated through an array of other factors and that hormones do not directly have specific biological or behavioral effects (1995: 595). As Oudshoorn has demonstrated, designating hormones as male or female essentializes differ-ences in male and female bodies. Scientific logic assumes that that male hor-mones create males and female hormones create females (Oudshoorn, 1994, 1991; see also Fausto-Sterling, 2000a). According to Oudshoorn, scientists are "actively constructing reality rather than discovering reality" (2001: 202). Often-times, the scientists themselves are aware of this situation. Blackless, Derryck, Charuvastra, Fausto-Sterling, Lauzanne, & Lee state that:

> Biologists and medical scientists recognize, of course, that absolute dimor-phism is a Platonic ideal not actually achieved in the natural world. None-theless, the normative nature of medical science uses as an assumption, the proposition that for each sex there is a single, correct developmental pathway. Medical scientists, therefore, define as abnormal any deviation from bio-medically distributed genitalia or chromosomal composition.
>
> (2000: 151)

Biologists and medical scientists are aware of the ideological underpinnings of their so-called hard sciences, yet, they still buy into the normative structures and modes of differentiation.

Coming from all fields of knowledge production, feminists started to be con-cerned with the relationship between biology and politics and started viewing and interrogating biological knowledge production from a critical feminist per-spective (Rees, 2007). Emily Martin, anthropologist, for example, studied the power of metaphor in modern medicine in her book *The Woman in the Body*

(1989). Martin interviewed more than 150 women from different backgrounds of ethnicity, age, and class. She demonstrated that menstruation is constructed differently throughout time and space and on the basis of race and class. Moreover, she reveals physicians' conceptualization of women's bodies and their reproductive abilities as value-laden in respect to the sexist and racist rhetoric of society. Martin also demonstrated, however, that women resist the mechanical metaphors to which medical practice and expertise subjects their bodies. In an article on *The Egg and the Sperm* (1996) Martin shows how stereotypical narratives of male and female behavior are used to describe reproductive processes on the level of egg and sperm. Donna Haraway, prominent feminist historian of biology and science detected assumptions about gender in the studies of the behavior of primates (1992). Her research reveals that interpretations of animal behavior are intrinsically connected to the socio-political background of the researcher. The representations, which follow from studies of animal behavior, reflect the researcher's perspective upon human behavior and not those of animals. She analyzed 'human nature' as an effect of power, which is rearticulated and re-produced permanently by the negotiation of the boundaries between human and animal, body and machine (1991, 1997). Evelyn Fox-Keller, originally trained as a physicist, deconstructed the claim to objectivity in the natural sciences by showing how cultural stereotypes of masculinity play into the construction of the scientist and *his* profession (1983). Her research focuses on the deconstruction of narratives in genetics and the uses of metaphors, models, and the notion of development in biological narratives (2000, 2002). Her research demonstrates that biological concepts are culturally located and shaped by complex social forces. Ruth Bleier, a neurophysiologist, criticizes Darwinism and socio-biology, and analyzes the interconnection and interdependencies of the natural sciences with power structures such as patriarchy, capitalism, and colonialism (2002). She mainly analyzes the production of biological entities such as genes as independent from environmental influences. Even minute biological units, she reveals, interact with the surroundings. Lynda Birke, feminist biologist, has criticized the abstract manner in which anatomy is depicted and how the body is thereby produced as having a surface and an inner space (2002). Birke describes how our inner and outer spaces are constructed as diametrically opposite and distinct, and how this distinction is communicated by gendered language. She argues that biology should be seen in the light of other disciplines from which it recruits its vocabulary: cybernetics, statistics, the arts, and facilities of abstraction, the non-spiritual perception and conception of the world (e.g., our skin as the limit, frame, and enclosure). She powerfully reveals the androcentrism and racism in the discipline of biology by interrogating the terminology from socio-political spheres, which reaffirm the status quo it relies upon. Barbara Duden, German medical historian (1997) read the rhetoric of the cold war in the descriptions of the immune system. The motherly body has been detected as one of the most powerful stereotypes of the female body. Reproduction has, of course, been one of the most interesting topics for feminists concerned with the construction of bodies. Ruth Hubbard, feminist biologist, for instance, has investigated the relationship between bioethics, new reproductive technologies, and genetic engineering (Hubbard, 1990; see also Sawicki, 1991).

Anne Fausto-Sterling pointed out in *Myths of Gender* (1985) that it is often extremely difficult to unravel arguments about the way in which biology is supposed to determine human behavior. This becomes especially clear in intersexualization where behavior and identity are at stake. Several feminist historians of science and critical biologists have therefore powerfully shown how sociopolitical circumstances influence 'biology talk' about the body and consequently sex(ual) difference. All these very different accounts show that there is no neutrality in the context, in which knowledge production happens or the intentions that motivate biological research into the body. And they demonstrate that there is no timeless and unchanging 'nature' of the body.

The most prominent feature in intersexualization is the scientific quest to differentiate males from females. The specifics of the details constantly change, but the quest remains the same. It easily adapts itself to the newest scientific developments. Every new discovery about the body's makeup can be interpreted in the coordinates of maleness and femaleness. Body historians of various shades and perspectives therefore started approaching the images of biological 'facts' as embodying 'reality.' These biological facts embody reality, as critical intersex scholar Iain Morland states, precisely through "refolding the interrelated discourses of knowing and telling. The image is real not because it offers a transparent window to the world, but because it actively contributes to the world's materiality" (2001: 542). Moreover, any investigation into biological facts, like chromosomes, or hormones, first represents the techniques of microscopy, molecular biology, photography, printing, and bookbinding—and then the shape of the chromosomes, of the substances of hormones. Therefore, the body's reality is first and foremost the representation of technologies and their limits. The body's materiality, as a result, can be seen as emerging through the materiality of technologies and their boundaries with regards to representation. Furthermore, feminist research in technologies has immensely contributed to a critical perspective on the application of technologies as part of male supremacy and patriarchal power (Rothschild, 1983; Silverstone & Hirsch, 1992; Wajcman, 1996). With regards to technological development and intersexualization Suzanne Kessler states that:

> the equation of gender with genitals could only have emerged in an age when medical science can create genitals that appear to be normal and to function adequately and an emphasis on the good phallus above all else could only have emerged in a culture that has rigid aesthetic and performance criteria for what constitutes maleness.

> (1998: 26)

Consequently, the critical stance towards technology and the advances in medical surgical techniques have also been considered and revealed as phallocratic. Noreen Giffney and Michael O'Rourke have recently identified a new biopolitical shift in research into intersexuality which they describe as "an attempt to control, discipline, render vulnerable and manageable the intersex body, an attempt to make the edgy body less troubling, to keep it before the law" (2009: xi). These attempts to keep the intersex body at bay and to make it less threating are apparent in a wide variety of approaches. It happens even in those accounts that attempt to provide the intersexualized body with more space for interpretation and treatment.

The Five Sexes

Of course it is no surprise that mainstream scientific research into sex differentiation would assume *the fact of* dimorphic sexual differentiation as its premise. This assumption, of course also motivates feminist biological research into intersexualization. Most prominently biologist Anne Fausto-Sterling, who specializes in evolutionary biology, endocrinology, and the neurosciences which makes her account on intersexuality utmost comprehensive from a biological perspective. Her 1993 provocative article "The Five Sexes" more or less initiated the intersex movement in the USA. Shortly after its publication, activist Cheryl Chase responded to Fausto-Sterling's article, and her reply received letters from intersexualized people around the world. Encouraged by the outpouring of supporters, Chase founded the Intersex Society of North America (ISNA). In 1994, the first issue of the newsletter *Hermaphrodites with Attitude* (ISNA 1994) was published, and in January 1995, a support group was founded. ISNA went online in January 1996.

Despite the seemingly radical intervention of Fausto-Sterling's suggestion that there are five sexes, the logic of her proposal fits squarely into the same phenomenon and repeats the same processes of intersexualization that John Money and Robert Stoller inaugurated (see Chapters 1 and 2 in this volume). Her work still is immensely important for its de-pathologizing approach to intersexuality. Nevertheless, an evaluation of her approach is critical for critical sex studies and the historicization of the (inter)sexualized body. Fausto-Sterling's article begins with an anecdote from 1843 in which the right to vote was at stake—the body of the person in question had to be examined by a physician. The result of the examination was that the determination of the body was prevented by its complexity. The example, Fausto-Sterling uses, shows that the motif of designating one sex or the other to a body comes from another source than biology—in this case historical laws of election. The author argues that the state and legal systems are in defiance of nature if they uphold a "two-party sexual system" because "biologically speaking, there are many gradations running from female to male" (1993: 21). Fausto-Sterling builds up her argument of (at least) five sexes, which deserve to be "considered additional sexes" (21). Her article continues to cite John Money on the frequency of intersex births, refers to the mythical origin of the word hermaphrodite, lists other instances of documentation of 'mixed' sexes and frames medical intervention as a biopolitical example in the Foucauldian sense. Although Fausto-Sterling is clearly opposed to surgical intervention in intersexualizations, assumptions of sexual dimorphism are to be found in her work. "The Five Sexes" controversially advocates we abandon the two-sex system in favor of recognizing five sexes: two extreme states of male and female and three middle states that cover what we would think of as hermaphrodite/intersex. The three middle or intermediate states that Fausto-Sterling proposes, are the following:

> the so-called true-hermaphrodites, whom I call herms, who possess one testis and one ovary (the sperm-and-egg-producing vessels, or gonads); the male pseudohermaphrodite (the "merms", who have testes and some aspects of the female genitalia but no ovaries; and the female pseudohermaphrodites (the "ferms", who have ovaries and some aspects of the male genitalia but lack testes).
>
> (1993: 21)

These three—herm, merm, and ferm—"deserve to be considered additional sexes each in its own right" (21). Sex is a "vast, infinitely malleable continuum that defies the constraints of even five categories" (21). Fausto-Sterling states and offers three intersexualized bodies as 'additional' to the 'original' or 'generic' categories. This notion of an original (upon which additions can be made) occurs in the vast majority of intersex narratives—not just those by psychologists and physicians, but also those by feminist scientists writing about intersexuality.[3]

With her proposal, Fausto-Sterling repeats the nineteenth-century Victorian classificatory system that privileged the gonads as the salient factor determining sex as delineating true hermaphrodites and pseudo-hermaphrodites. Alice Dreger called this era the Age of Gonads and characterized this period as unified in its approach to hermaphroditism (1998). Ulrike Klöppel, however, demonstrated that the unifying term Age of Gonads disregards the multiple and controversial approaches by numerous researchers at the time (2010). Dreger's conceptualization demonstrates that different categorization systems are repeatedly called upon to make hermaphroditism/intersexuality intelligible for a wider audience. Moreover, it proves that the Age of the Gonads is not a historical phenomenon limited to a specific time frame it rather reoccurs at certain points in time. In fact, it reoccurs in a feminist biological account more than a hundred years later.

Fausto-Sterling in her groundbreaking article, interestingly proposes a sex difference, which is based on reproductive capabilities or possibilities in an era that has just started to remove reproduction from the human body into a petri dish. Just to remind the reader, this, interestingly, contradicts the Money protocol from 40 years before. According to Money, sex assignment should not be based on the gonads and reproductive ability. Money, in fact, dismissed reproductive capability as the decisive factor for surgical intervention in favor of the adaptability of the person to one or the other gender role, based on the appearance of the genitalia (see Chapter 1 in this volume and Money et al.'s Recommendations Concerning Sex Assignment 1955). The Victorian classification that Fausto-Sterling champions, reiterates the reproductive imperative. In doing so, it unfortunately also ensnares hermaphroditism/intersexuality inside the abject status it has been assigned in the heteronormative regime. Even though Fausto-Sterling acknowledges that intersex* bodies raise the specter of homosexuality and salutes this, her argumentative strategy of using reproductive organs as decisive factor however, repeats and reiterates heteronormative instances. Whenever sex is brought up in combination with the body's ability to reproduce, heteronormativity is implicit. It makes us forget that only a minimal proportion of heterosexual sex is actually aimed at reproduction. Heterosexual sex often does not involve reproductive organs, yet, it is always assumed that genitals are used in the heteronormative complementary way. Gender role is what is at stake, when it comes to gender assignment as Money has clearly demonstrated when he disregarded reproductive ability as a decisive factor in intersexualization (see Chapter 1 in this volume).

Suzanne Kessler has also offered a critique of Fausto-Sterling's proposal and exposes the problem of a gender-concept that always implies sex. Kessler admits that "validating alternative genitals validates alternative 'sexes' and damages the privilege of 'female' and 'male'" (1998: 90). "But," she continues, "the issue is gender and not 'sex,'" meaning that the bodies of the intersex* persons are not the problem, it is them not fitting into one or the other gender role and thereby questioning the exclusivity of the binary system of gender. The limitation with Fausto-Sterling's proposal of the "Five Sexes" is, according to Kessler:

> that legitimizing other sets of genitals still gives genitals (albeit in more than two forms) primary signifying status and ignores the fact that in the everyday world gender attributes are made without access to genital inspection. There is no sex, only gender because gender performance is the primary in everyday life. This gender is performed regardless of the flesh's configuration under the clothes.
>
> (1998: 90)

Even though Kessler incorrectly identifies genitals and not gonads as the body part Fausto-Sterling focuses on, her critique that sex is always already gender is valid. Also her argument that giving signifying status to a specific part of the body—such as the genitals, or the reproductive organs is to imply that sex, or the materiality of the body has the power of determining and constituting gender. This also testifies to the synecdochal logics which is always implicit in intersexualization, meaning a part of the body stands in for the whole of the body and its meaning as regards the search for *true sex* (see Chapter 1 in this volume).

Fausto-Sterling evokes Foucault's concept of bio-power to argue in accordance with him that the medical establishment has fostered "the complete erasure of any form of embodied sex that does not conform to a male-female, heterosexual pattern" (1993: 23). She continues "ironically, a more sophisticated knowledge of the complexity of sexual systems has led to the repression of such intricacy" (23) and the "knowledge developed in biochemistry, embryology, endocrinology, psychology and surgery has enabled physicians to control the very sex of the human body" (24). Therefore, she reads Foucault's conceptualization of knowledge production in line with Jana Sawicki, a renowned Foucault scholar, who argues that "deviancy is controlled and norms are established through the very process of identifying the deviant as such, then observing it, further classifying it, monitoring and 'treating' it" (1991: 39). The biological knowledge production of the two sex system, sophisticated or not, produces deviant entities to consolidate the norm. Because gender is always already there bifurcation of every detail relating to sex to be discovered will be interpreted as dimorphic. Fausto-Sterling rightly asserts that the "medical accomplishment" should not be read as "progress" but as a mode of discipline, because "hermaphrodites have unruly bodies" (24) who are not just controlled by surgery—they are in fact erased as such.

Five Sexes Revisited

In "The Five Sexes Revisited," seven years later Fausto-Sterling argues in reference to the medical ethicist Laurence B. McCullough, that the various forms of intersexuality should be defined as normal (2000b). Nevertheless, she does not question the necessity of sex-assignment itself, which she suggests, in line with McCullough, should be done "on the basis of the probability that the child's particular condition will lead to the formation of a particular gender identity" (21). She cautions, however, that:

> at the same time, though, practitioners ought to be humble enough to recognize that as the child grows, he or she may reject the assignment—and they should be wise enough to listen to what the child has to say.
>
> (21)

Fausto-Sterling concedes with this view, stating that surgery should only happen if it does not pose any medical danger to the child. Moreover, Fausto-Sterling this time reports of intersex individuals themselves. However, genital surgeries often leave lifelong negative impacts on the individual, reducing sexual sensitivity and causing traumatic experiences. Fausto-Sterling therefore, recommends therapy, not surgery, and she mandates that physicians submit to ethical guidelines concerning patients' rights and informed consent. These include informing the parents, abandoning newborn surgery, and minimizing irreversible assignments.

Whereas in the article from 1993 Fausto-Sterling was still talking about the continuum of sex, she now states that "sex and gender are best conceptualized as points in a multidimensional space" (2000b: 22). And as a biologist she knows about recent developments in biology and explains the different biological levels which are employed in distinguishing sex:

> For some time, experts on gender development have distinguished between sex at the genetic level and at the cellular level (sex-specific gene expression,

X and Y chromosomes); at the hormonal level (in the fetus, during childhood and after puberty); and at the anatomical level (genitals and secondary sexual characteristics). Gender identity presumably emerges from all of those corporeal aspects via some poorly understood interaction with environment and experience. What has become increasingly clear is that one can find levels of masculinity and femininity in almost every possible permutation.

(22)

Therefore, the medical community is adapting to the lived reality of individuals and beginning to accept that sex is a complex issue. Fausto-Sterling concludes that "the medical and scientific communities have yet to adopt a language that is capable of describing such diversity" (22). Thereby Fausto-Sterling hints at one of the main problems in intersexualization: the insufficiency of—not just— biological language for the description of non-conforming bodies. Yet, again: in the end it all circles around the issue of gender and not sex. Fausto-Sterling's conclusion testifies to this, she states: "there are and will continue to be highly masculine people out there; it's just that some of them are women. And some of the most feminine people I know happen to be men" (23). It always boils down to who behaves appropriately to heterorelational and -normative standards. "Cultural genitals" (Kessler, 1998) most of the time cover "anatomical genitals"— why are they still needed for distinguishing social roles? To repeat Foucault's question from his introduction to Herculine Barbin's diaries: "Do we truly need a true sex?" (1980: vii). It seems, for sure, that we need to give a sex to the body, even though the truth has not much to do with it—at least not in the case of intersexualization.

Anne Fausto-Sterling's *Sexing the Body* (2000a) discusses a 1988 Olympics case similar to Caster Semenya's (see Introduction in this volume). The International Olympic Committee (IOC) tested Maria Patiño to determine whether she was properly assigned to the women's division. Chromosonal tests discovered Patiño had a Y chromosome and diagnosed her with Androgen Insensitivity Syndrome (AIS), which is classified as an intersex condition. She was banned from the Olympics and only later rejoined after being reinstated by the International Amateur Athletic Federation (IAAF). Patiño went down in the annals of professional international sports competitions as "the first woman ever to challenge sex testing for female athletes" (2). Gender-testing was mandatory for female athletes at the Olympics from then, but dropped in 1999 because the IAAF realized that it was a sexist practice. In her analysis of this particular case as well as in the complete monograph, Fausto-Sterling addresses the techniques, technologies, discoveries, experts, and other details of scientific discourses concerning the sexed body in current biological knowledge production.[4] She effectively illustrates the constitution of the powerful bio-medical discourses and investigations and reveals how social paradigms disguise themselves in biological terminologies and how heteronormativity serves sex_gender assignment procedures. Fausto-Sterling exposes the degree to which socio-political ideologies shape and produce 'pure' science, resulting occasionally in lives ruined by the violation of the body and integrity of intersexualized children. She rejects traditional classification of intersexualized people as male, female, and true or pseudo-hermaphrodites and in doing so, provides a basis for de-pathologizing those who are considered to fall within the intersex* umbrella (50). However, her reliance upon recent developments in the field prevents her from completely escaping the tendency of bio-medicine to pathologize intersexualized individuals (51–53). Fausto-Sterling compiles psychological profiles of intersex-identified children raised as male or female. The comments she chose to suggest they turned out to be psychologically either healthy or unhealthy adults almost exclusively refer to the (heterosexual) marital status of the person (96–100; 102–106). However, whereas before she argued that some men are more feminine than some women, now she argues that "ultimately, perhaps, concepts of masculinity and femininity might overlap so completely as to render the very notion of gender difference irrelevant" (101). In *Sexing the Body*, she has reconsidered her proposal and revised her argument of using the "discrete categories such as herm, merm, and ferm, even tongue in cheek" (110). Instead she admits more clearly to the dominance of gender in the debate as well as to the heterosexual matrix that influences every single term in intersexualization. She states that the "debates over intersexuality are inextricable form those over homosexuality" (112) and that sexual performance needs to be re-considered in intersexualization, especially in the light of "unusual genitals" (113). "Perhaps" she argues "we will come to view such children as especially blessed or lucky" because they might become the "most desirable of all possible mates, able to pleasure their partners in a variety of ways" (113). Even though Fausto-Sterling subjects her biological study to biological paradigms and terminology, she is able to argue for a utopian society in which sexual performance is conceptualized as ingenuity and not as heteronormative compliancy. Sexual difference however, is in biological parlance or in terms of a philosophical framing not released and intersex remains 'the other' to the gender_sex binary.

Feminist Materialism—a Way Out of Intersexualization?

Intersexualization is based on a belief of sexual difference as binary. The gender-concept as I have shown, was in fact invented in intersexualization (see Chapters 1 and 2). The distinction between sex and gender originates in the pathologization of discrepancies between the two—both always thought of as exclusively binary. Can sexual difference as a concept be freed from the dichotomy—does it have to be freed from the binary in order to de-pathologize intersexuality—in effect to reverse intersexualization? In search for concepts and approaches to sexual difference not necessarily bifurcated one ends up with new feminist materialism. New feminist materialists are greatly influenced by the philosophy of Gilles Deleuze and Félix Guattari (e.g., Buchanan & Colebrook, 2000). They believe that we do not need the gender-concept in order to fight discrimination. Feminist materialists go back to Simone de Beauvoir who claimed in *Le deuxième Sexe* as early as 1949 that one is not born, but rather becomes woman—all completely without gender as an analytical category. In new feminist materialist philosophies sexual difference per se is not understood as discriminating—it is understood as simply differing. In an interview with Rosi Braidotti key concepts in feminist materialism are named as "the sexualized nature and the radical immanence of power relations and their effects upon the world" (Dolphijn & van der Tuin, 2012: 22). Sexual difference, for new material feminists is neither given nor culturally constructed: they conceptualize sexual difference and reproduction as emerging from "parallel processes of transmission of information among diverging microbial bodies" (Parisi, 2004a: 22). The focus of new feminist materialist thinkers therefore is shifted from individual human bodies to much smaller entities and interactions and events between them. This enables a crossing of material and discursive boundaries and philosophical as well as biological concepts alike. New material feminists conceive of matter as an active agent. They do this in order to ensure an understanding of the mutual articulation of matter and meaning. Towards the conclusion of this chapter, I merely wish to provide the reader with some tags that I view as promising in new materialist feminist thinking in regards to an intervention into intersexualization, albeit I cannot give a full account on this new strand of theorizing due to limited space.[5]

In *A Thousand Plateaus,* Deleuze and Guattari argue that sexuality "is badly explained by the binary organization of the sexes, and just as badly by a bisexual organization within each sex. Sexuality brings into play too great a diversity of conjugated becomings; these are like n sexes" (2004: 307). They foster the notion of unlimited difference and the impossibility of applying a system (of two, three, or any number) less than n. This n-th degree of the different possible materializations of a sex-gender-sexuality system dissolves all categories. 'Humanity' or 'human nature' as Deleuze and Guattari argue, is an underlying, sexless essence that is then boxed into different bodies and identities. In Deleuze and Guattari, we also find the notion that embodiedness of the subject is a form of bodily materiality, albeit not in terms of a biological essence. For them the body is the complex interplay of exceedingly constructed social and symbolic forces.

Feminist philosopher Elisabeth Grosz first provided in 1994 an analysis of the sexed body as a living and experiencing entity in *Volatile Bodies*. Grosz argues that the body is not a brute, passive, or, inert object merely inscribed by social forces, but rather that it is created through already established social systems of representation, meaning, and signification. For corporeal feminists such as Grosz, "the body can be seen as the crucial term, the site of contestation, in a series of economic, political, sexual and intellectual struggles" (19). Her work has caused feminist scholars to engage with the possibilities of reclaiming phallocentric knowledge production in the natural sciences. Australian feminist thinker Moira Gatens, critical of purely constructivist approaches, can be interpreted as saying that "the idea that the world is constructed through language merely repeats a centuries-old privilege of the formal and logical over the material" (Colebrook on Gatens, 2008). New feminist materialists have in recent years stressed that there cannot be simple material notions of life. This new approach to the body is marked by the call for thinking the body beyond the problem of representation because there is no original body that could be represented. These new approaches to the body stress difference on all levels and are therefore of specific interest as regards intersexualization and accompanying binary definitions of sexual difference.

New material feminists argue that evolutionary change occurs not in a mechanical manner as has been widely believed. Claire Colebrook, cultural theorist and Deuleuzian scholar, states that "if there is an evolutionary imperative at the heart of our emotional and rational tendencies, then this is not because of hardwiring but because the brain is adaptive and dynamic" (2008: 53). Colebrook, who also works with the ideas of French philosopher Henri Bergson, sees life as the production of difference and the subsequent responsive and dynamic relation among differences. Life, according to her, should also "not be conceived as matter, for matter is nothing more than the relatively stable form taken on by a life that is truly and fundamentally a potentiality for change" (54). What follows for Colebrook is a "material politics," which consists of two movements: one is the recognition of how life has unfolded historically to produce the relations we are living in, and the second is that the freedom from these relations can only be achieved through recognition of our materiality. With regards to the production of a dichotomous system of the sexes and therefore intersexualization this means that there is a need to acknowledge the material history of how this has been produced but also and importantly the "refutation that such a history is inevitable" (64). For Colebrook it follows that "our biology can and should be lived otherwise" (65). Biology, here understood as the material is acknowledged beyond a constructivism versus essentialism debate. Authors Stacy Alaimo and Susan Hekman (2008) draw attention to the "lived material bodies and evolving corporeal practices" (3) and stress that, although "language structures how we apprehend the ontological, it doesn't constitute it" (98). In this logic, an ontology of difference enables engineering a "body politics for feminism that highlights differential changes rather than locating sexual difference in a given grid of power" (Parisi, 2004b: 81).

Contradicting patriarchal, sexist, and deterministic conceptualization of the human body as binary and complementary, new feminist materialists argue that "the matter of the human body is just that which can *give itself* any form whatever" (Colebrook, 2008: 79). Luciana Parisi's material feminism recognizes modern technology. She states that "if molecular biotechnology is already detaching femininity from the imperative of sexual reproduction and genetic sex then why would a notion of femininity be relevant to the body politics?" (2004b: 81). If bodies are no longer divided by the "reproductive imperative," the concept of sexual identity as being based on dimorphic sexual difference, or in Robert Stoller's sense, gender identity, becomes obsolete. And not just this becomes obsolete but according to Rosi Braidotti, who describes Deleuze's and Guattari's approach to identity, the "very concept of identity, with the inbuilt logic of recognition of sameness and dualistic relocation of otherness, which has been operational since Plato's time" (Braidotti, 2005) will be superceded. Braidotti touches upon the tendency towards non-dialecticism in feminist materialism and invokes a basic reorganization of Western thinking. Braidotti reasons that with a Deleuzian approach, an ethics of sexual difference will replace metaphysics altogether.

With regards to intersexualization this is, of course, a foundational claim because sexual difference organized as sexual dimorphism is the phantasm, which produces the violent effects on intersexualized bodies and identities. With Grosz, Colebrook argues for a new:

> thinking of the body neither as a pre-representational surplus nor as a deter-mining essence. The question is not one of how the sexes are differentiated, but rather: are there different modalities of sexual differentiation due to the specificity of different bodies?
>
> (Colebrook, 2000 [emphasis in original]; Grosz, 1994: 189)

In the light of my argument that the binary sex-gender distinction was invented in intersexualization, this approach seems very promising for future interventions into intersexualization.

Sexual difference, as implying sexual dimorphism and biological determinism, I argue, still needs to be convincingly historicized by correlating the constructions of other differential modes of human bodies; the interplay of social and symbolic forces which, in their interplay, produce multiple vectors of power and hierarchy. Several modes of differentiation take part in the construction of bio-medical knowledge. The categories of white, black, poor, rich, ablebodied, old, and young, for example, play a significant role in enabling biology to speak at all. The language that mediates 'scientific experiments' is a reverse and circulating process of power. Analogies of race and sex(ual) difference are mediated by evolutionary tropes such as *maturity, development, degeneration* and so forth, as I elaborate in Chapter 6 in this volume. Numerous concepts of the sex(ualiz)ed_ gendered body and the 'racially' different body, i.e., the racialized body, could not be made intelligible without a circular system of references and metaphorical analogies. In de-constructing sex(ual) difference the symbolic order, power relations, technologies, psychology, linguistics, and daily-life-interactions and the meaning of matter cannot be left aside.

Conclusion

Feminists have eagerly taken on the distinction between sex and gender from Robert Stoller's publication in 1968. The gender-concept has since provided feminists with numerous possibilities to historicize and fight gender discrimination. However, the implicit pathologization of non-conforming bodies in the distinction between sex and gender reoccurs on several occasions and from time to time prevents the initial aims of freeing gender from sex. Intersexualization prefigures research into sex, gender, and sexuality not just by non-feminist psychologists, psychoanalysts, and bio-medical researchers but also by feminist researchers. However, especially in critical feminist biology, convincing approaches to the de-essentialization of sex(ual) difference were developed. Yet, when it comes to intersexuality, problems in conceptualization remain. In the light of my argument that the gender-concept was invented in intersexualization, I contend that the distinction of sex and gender is intrinsically connected to the continuing effort to align the two categories through surgical and/or therapeutic intervention. Moreover, with regards to feminist and queer theoretical engagement with the gender-concept, the fallacy of the easy deconstructability of sex always re-enters through the back door. The heteronormative bias, which is implicated in the binary constructions of sex and gender *and* their distinction, has made theories that avoid reinstalling the pathology of sex-gender-incongruence difficult, yet not impossible.

The politicization and historicization of knowledge production in the discipline of biology from a feminist perspective was clearly enabled by the distinction between sex and gender. The work by critical feminist biologists, here especially Anne Fausto-Sterling, has been indispensable and has powerfully deconstructed the alleged 'objectivity' of the knowledge productions in bio-medicine. A number of inherently patriarchal and sexist tautological arguments can however be dismantled. Sex(ual) difference as a dimorphic biological imperative has been weakened by the critical feminist biologists' work. However, the separation of sex and gender fostered a division in academia that reinstalled the split. And I am much inclined to accept Toril Moi's statement that a useful understanding of the body and subjectivity is not to be achieved if we remain in the notion of a biological body and a psychological subjectivity (1999). Of course, theorists in the Butlerian tradition of handling sex as already gendered have provided accounts that can overcome the distinction, but in regard to intersexualization the distinction usually prevails. I have shown with the example of Fausto-Sterling that feminists who engaged with intersexualization while choosing to use the distinction between sex and gender are unwillingly complicit. Dichotomously thought sex(ual) difference enters the narratives on the body even though researchers try to avoid reinstalling binary heteronormative concepts. No matter how many categories one choses to include into the calculation, the dichotomy remains intact. Although gender was able do an immense work for feminists interrogating the interconnectedness of gender with other categories such as race, ethnicity, age, and so on, sex is always difficult to not imply in the use of the category of gender. I very briefly touched upon the new approaches by materialist feminist to sexual difference and the knowledge productions in bio-medicine. I argued that their reconfiguration of materiality within a philosophical framework is highly promising also in regards to intervention into intersexualization.

Notes

1. Stoller laid the ground for later theories of prenatal hormonal influences on the brain, thereby preconceiving the so-called masculine or feminine development of the psyche as later developed by Milton Diamond (Diamond & Sigmundson, 1997; Diamond, 1999).

2. Criticism of the unified notion of the one-sex model as implied by Laqueur has been voiced by a number of authors (e.g., Daston & Park, 1995, Schleiner, 2000), who stressed that they do not see it as simply and coherently represented as Laqueur does. Moreover, Laqueur's account is problematic with regard to his installation of the *two-gender system* as ahistorical. (Daston, 1991; Daston &, Park, 1995; Epstein, 1990; Jones, & Stallybrass, 1991; Long, 1999). Even though Lacqueur manages to historicize sex, he ontologizes gender by assigning the category an ahistorical presence thereby reaffirming the distinction. However, his account has powerfully shown that 'biological' explanations of the body are deeply rooted in the political, cultural, and social orders of the word.

3. Freud, however, assumed that intersex (bisexuality in his terms) was the primary state from which the other sexes developed. Interestingly this narrative of 'unfinishedness' (see Chapters 1 and 6 in this volume) does feature on different occasions in intersexualization, it nevertheless is never used to argue for the naturalness or normality of intersex/hermaphroditism.

4. Fausto-Sterling uses *development systems theory* (DST) in this volume. This enables her to depict the body as biological but also tightly bound to environmental factors. DST is based on three basic principles: (1) Nature and nurture are indivisible. (2) Organisms are active processes (all throughout life). (3) No single academic discipline can tell the truth about human nature (and human sexuality) (Fausto-Sterling, 2000a: 235).

5. For a comprehensive account on the new materialist feminists see Iris van der Tuin 2008. Van der Tuin refers to scholars such as Claire Colebrook, Sarah Ahmed, and Karen Barad as materialist feminists.

Bibliography

Alaimo, S., & Hekman, S. (Eds.). (2008). *Material feminisms*. Bloomington: Indiana University Press.

Beauvoir, S. de (1949). *Le deuxième sexe*. Paris: Gallimard.

Birke, L. (1999). *Feminism and the biological body*. Edinburgh: Edinburgh University Press.

Birke, L. (2002). Anchoring the head: the disappearing (biological) body. In G. Bendelow , M. Carpenter, C. Vautier, & S. Williams (Eds.), *Gender, health and healing. The public/private divide* (pp. 34–48). London, New York: Routledge.

Blackless, M., Derryck, A., Charuvastra, A., Fausto-Sterling, A., Lauzanne, K., & Lee, E. (2000). How sexually dimorphic are we? Review and synthesis. *American Journal of Human Biology, 12*(2), 151–166.

Bleier, R. (2002). Sociobiology, biological determinism, and human behavior. In M. Wyer, M. Barbercheck, D. G. Cookmeyer, H. Ozturk, & M. Wayne (Eds.), *Women, science, and technology: A reader in feminist science studies* (pp. 175–193). London: Routledge.

Bordo, S. (1993). The body and the reproduction of femininity. In K. Conboy, N. Medina, & S. Stanbury (Eds.), *Writing on the body. Female embodiment and feminist theory* (pp. 90–110). New York: Columbia University Press.

Braidotti, R. (2005). Affirming the affirmative: On nomadic affectivity. *Rhizomes, 11/12*. Retrieved March 18, 2016 from www.rhizomes.net/issue11/braidotti.html.

Buchanan, I., & Colebrook, C. (Ed.). (2000). *Deleuze and feminist theory*. Edinburgh: Edinburgh University Press.

Butler, J. (1990). *Gender trouble. Feminism and the subversion of identity*. New York: Routledge.

Butler, J. (1993). *Bodies that matter: On the discursive limits of 'sex.'* New York: Routledge.

Campbell, J. (2000). *Arguing with the phallus: Feminist, queer and postcolonial theory: A psychoanalytic contribution.* London: Zed Books.

Colapinto, J. (2000). *As nature made him: The boy who was raised as a girl.* New York: Harper Collins.

Colebrook, C. (2000). From radical representations to corporeal becomings: The feminist philosophy of Lloyd, Grosz, and Gatens. *Hypatia, 15*(2), 67–93.

Colebrook, C. (2008). On not becoming man: The materialist politics of unactualized potential. In S. Alaimo and S. Hekman (Eds.), *Material* feminisms (pp. 52–83). Bloomington: Indiana University Press.

Daston, L. (1991). Marvelous facts and miraculous evidence in early modern Europe. *Critical Inquiry, 18*(1), 93–124.

Daston, L., & Park, K. (1995). The hermaphrodite and the orders of nature. Sexual ambiguity in early modern France. *GLQ: A Journal of Lesbian and Gay Studies, 1*(4), 419–438.

Deleuze, G., & Guatarri, F. (2004). *A thousand plateaus: Capitalism and schizophrenia.* London: Athlone.

Diamond, M. (1999). Pediatric management of ambiguous and traumatized genitalia. *The Journal of Urology, 162,* 1021–1028.

Diamond, M., & Sigmundson, H. K. (1997). Management of intersexuality. Guidelines for dealing with persons with ambiguous genitalia. *Archives of Pediatrics & Adolescent Medicine, 151,* 1046–1050.

Dolphijn, R., & Tuin, I. van der (2012). *New materialism: Interviews & cartographies.* Michigan: Open Humanities Press.

Dreger, A. D. (1998). *Hermaphrodites and the medical invention of sex.* London: Harvard University Press.

Duden, B. (1997). Das "System" unter der Haut. Anmerkungen zum körpergeschichtlichen Bruch der 1990er Jahre. *ÖZG, 8*(2), 260–273.

Elias, N. (1978). *The History of Manners. The Civilizing Process: Volume I.* New York: Pantheon Books.

Elias, N. (1982). *Power and civility. The civilizing process: Volume II.* New York: Pantheon Books.

Epstein, J. (1990). Either/or—neither/both: Sexual ambiguity and the ideology of gender. *Genders, 7,* 99–142.

Fausto-Sterling, A. (1985). *Myths of gender. Biological theories about women and men.* New York: Basic Books.

Fausto-Sterling, A. (1993). The five sexes: Why male and female are not enough. *The Sciences, 33*(2), 20–25.

Fausto-Sterling, A. (2000a). *Sexing the body. Gender politics and the construction of sexuality.* New York: Basic Books.

Fausto-Sterling, A. (2000b). The five sexes, revisited. *The Sciences, 40*(4), 18–23.

Feher, M., Naddaff, R., & Tazi, N. (1989). *Fragments of a history of the human body.* New York: Zone Books.

Flax, J. (1990). Postmodernism and gender relations in feminist theory. In L. Nicholson (Ed.), *Feminism/postmodernism* (pp. 39–62). New York, London: Routledge.

Foucault, M. (1965). *Madness and civilization: A history of insanity in the age of reason.* New York: Pantheon Books.

Foucault, M. (1977). *Discipline and punish: The birth of the prison.* New York: Vintage Books.

Foucault, M. (1978). *History of sexuality*. London: Penguin Books.

Foucault, M. (Ed.). (1980). *Herculine Barbin: Being the recently discovered memoirs of a nineteenth-century French hermaphrodite*. New York: Pantheon Books.

Foucault, M (2002). Die Geburt der Sozialmedizin. In Foucault, M. (Ed.), *Schriften in vier Bänden. Dits et Ecrits. 1970–1975* (pp. 272–297). Frankfurt am Main: Suhrkamp.

Fox-Keller, E. (1983). Gender and science. In S. Harding and B. Hintikka (Eds.), *Discovering reality. Feminist perspectives on epistemology, metaphysics, methodology, and philosophy of science* (pp. 187–206). Dordrecht, Boston, London: D. Reidel Publishing Company.

Fox-Keller, E. (2000). *The century of the gene*. London: Harvard University Press.

Fox-Keller, E. (2002). *Making sense of life: Explaining biological development with models, metaphors and machines*. London: Harvard University Press.

Gatens, M. (1988). Towards a feminist philosophy of the body. In B. Caine, E. A. Grosz, & M. de Lepervanche (Eds.), *Crossing boundaries: Feminism and the critique of knowledges* (pp. 59–70). Sydney: Allen and Unwin.

Gatens, M. (1991). *Feminism and philosophy: Perspectives on difference and equality*. Cambridge: Polity Press.

Giffney, N., & O'Rourke, M. (2009). Series editors' preface: Intersex trouble; or, how to bring your kids up intersex (pp. ix–xii). In M. Holmes (Ed.), *Critical intersex*. Farnham: Ashgate.

Goldschmidt, R. (1916). Experimental intersexuality and the sex-problem. *The American Naturalist, 50*(600), 705–718.

Grosz, E. (1994). *Volatile bodies: Towards a corporeal feminism*. Bloomington, Indianapolis: Indiana University Press.

Haraway, D. (1991). *Simians, cyborgs, and women: The re-invention of nature*. London: Routledge.

Haraway, D. (1992). *Primate visions. Gender, race, and nature in the world of modern science*. New York, London: Routledge.

Haraway, D. (1997). *Modest_witness@second_millennium.femaleman©meets_oncomouseTM: Feminism and technoscience*. New York: Routledge.

Hubbard, R. (1990). *The politics of women's biology*. Piscataway, New Jersey: Rutgers University Press.

ISNA. (1994). *Hermaphrodites with Attitude, n.1.*

Jones, A. R. and Stallybrass, P. (1993). Fetishizing gender: Constructing the hermaphrodite in Renaissance Europe. In J. Epstein, and K. Straub (Eds.), *Body guards: The cultural politics of gender ambiguity* (pp. 80–111). New York: Routledge.

Kessler, S. (1998). *Lessons from the intersexed*. London: Rutgers University Press.

Klöppel, U. (2010). *XX0XY ungelöst: Die medizinisch-psychologische Problematisierung uneindeutigen Geschlechts und Trans/Formierung der Kategorie Geschlecht von der Zeit der Aufklärung bis in die Gegenwart*. Bielefeld: transcript Verlag.

Laqueur, T. (1990). *Making sex. Body and gender from the Greeks to Freud*. London: Harvard University Press.

Latour, B., & Woolgar, S. (1979). *Laboratory life: The construction of scientific facts*. Princeton University Press.

Levin, D., & Solomon, G. F. (1990). The discursive formation of the body in the history of medicine. *Journal of Medicine and Philosophy, 15*(5), 515–537.

Long, K. P. (1999). Sexual dissonance: Early modern scientific accounts of hermaphrodites. In P. Platt, (Ed.), *Wonders, marvels, and monsters in early modern culture* (145–163). London: Routledge.

Lorenz, M. (2000). *Leibhaftige Vergangenheit. Einfuehrung in die Koerpergeschichte*. Berlin: edition diskord.

Martin, E. (1989). *The woman in the body: A cultural analysis of reproduction*. Boston: Beacon Press.

Martin, E. (1996). The egg and the sperm: How science has constructed a romance based on stereotypical male-female roles. In E. Fox-Keller and H. E. Longino (Eds.), *Feminism and science* (pp. 103–120). Oxford, New York: Oxford University Press.

Mauss, M. (1936). *Sociologie et anthropologie*. Paris: PUF.

Merlau-Ponty, M. (1962). *Phenomenology of perception*. London: Routledge.

Moi, T. (1999). *What is a woman? And other essays*. Oxford: Oxford University Press.

Money, J., Hampson, J. G., & Hampson, J. L. (1955). Hermaphroditism: Recommendations concerning assignment of sex, change of sex, and psychologic management. *Bulletin of the Johns Hopkins Hospital, 97*(4), 284–300.

Morland, I. (2001). Is intersexuality real? *Textual Practice, 15*(3), 527–547.

Oakley, A. (2015 [1972]). *Sex, gender and society*. London: Temple Smith.

Oudshoorn, N. (1991). *The making of the hormonal body: A contextual history of the study of sex hormones*. Enschede: Druckerij Alfa.

Oudshoorn, N. (1994). *Beyond the natural body: An archeology of sex hormones*. London, New York: Routledge.

Oudshoorn, N. (2001). On bodies, technologies, and feminisms. In A. Creager, E. Lunbeck, & L. L. Schiebinger, (Eds.), *Feminisms in twentieth-century. Science, technology, and medicine* (pp. 199–213). Chicago: University of Chicago Press.

Parisi, L. (2004a). *Abstract sex: Philosophy, biotechnology and the mutations of desire*. London, New York: Continuum.

Parisi, L. (2004b). For a schizogenesis of sexual difference. *Journal for Politics, Gender and Culture*, *3*(1), 67–93.

Rees, A. (2007). Reflections on the field: Primatology, popular science and the politics of personhood. *Social Studies of Science*, *37*(6), 881–907.

Rothschild, J. (ed.) (1983). *Machina ex dea: Feminist perspectives on technology*. New York: Pergamon Press.

Sawicki, J. (1991). *Disciplining Foucault. Feminism, power and the body*. New York, London: Routledge.

Schleiner, W. (2000). Early modern controversies about the one-sex model. *Renaissance Quarterly, 53*(1), 180–191.

Silverstone, R., & Hirsch, E. (Eds.). (1992). *Consuming technologies. Media and information in domestic spaces*. London, New York: Routledge.

Stoff, H. (1999). Diskurse und Erfahrung. Ein Rückblick auf die Körpergeschichte der neunziger Jahre. *Max Planck Institut Hefte, 2*, 142–160.

Stoller, R. (1968). *Sex and gender. On the development of masculinity and femininity*. New York: Science House.

Stoller, R. (1985). *Presentations of gender*. New Haven, London: Yale University Press.

Tuin, I. v. d. (2008) Third wave materialism: New feminist epistemologies and the generation of European women's studies (unpublished Doctoral dissertation, Universiteit Utrecht).

Wajcman, J. (1996). *Feminism confronts technology*. Cambridge: Polity Press.

Worthman, C. (1995). Hormones, sex, and gender. *Annual Review of Anthropology, 24*, 593–617.

Part II
The Colony

4 Seeing, Hearing, Translating

In 1985, anthropologist Gilbert Herdt and psychoanalyst Robert Stoller, both from California, published an interview with Sakulambei, a Papua New Guinean Shaman. They conducted the interview together in 1979 when Stoller joined Herdt in Papua New Guinea. By then, Herdt had lived in Papua New Guinea on and off since 1975 and had conducted ethnographic research about a people that he referred to as the Sambia (a pseudonym) of the Papua New Guinean Highlands. Herdt has published widely on the societal organization, rites of passage, and gender relations among the Sambia (e.g., Herdt, 1981, 1982). Herdt invited Stoller, after having apprenticed with him on the issues of sexuality, erotics, and gender identity between 1975 and 1979 at UCLA. Herdt felt that it was necessary to bring the expertise of a clinician to his field site. Stoller went to Papua New Guinea for ten days specifically to help Herdt undertake research and to assist him in interviewing six individuals of the Sambia. One of them was Sakulambei, a Sambian Shaman whom Herdt suspected to be intersex. Herdt and Stoller therefore explicitly address intersexuality in "Sakulambei—A hermaphrodite's secret: An example of clinical ethnography" (Herdt & Stoller, 1985). The paper is a partial transcript of their interview with the shaman, and is republished in *Intimate Communications* (Herdt & Stoller, 1990), which includes six interviews with other Sambian "interpreters" as Herdt and Stoller call their interviewees (46).

For this first book, *Guardians of the Flutes* (1981), Gilbert Herdt worked predominantly with the Sambia of the Papua New Guinea Highlands; eroticism and sexuality were his central concerns. Aware of Robert Stoller's expertise in sexology, he contacted the psychologist/psychoanalyst from UCLA. Stoller subsequently invited Herdt to UCLA as a postdoctoral fellow. From 1978 to 1985, Herdt became a member of the Gender Identity Research Clinic at UCLA. In 1979, he received a Post-doctoral Certificate in Psychiatry at the Neuropsychiatry Institute. Since then, he has held positions at Stanford University, the University of Chicago, the University of Washington, and the University of Amsterdam. He has been Professor of Human Sexuality Studies and Anthropology, and Director of the National Sexuality Resource Center at San Francisco State University since 2002. He is currently an associate editor of the *Journal of Culture, Sexuality, and Health*; the *Journal of Men and Masculinities*; and *Transaction:*

Journal of Social Science and Modern Society. Robert Stoller was most famous for his work on intersexuality and transsexuality, but he has also published on transvestism, erotic imagery/pornography, sado-masochism as a pathological expression of sexuality, and homosexuality (see Chapter 2 in this volume).

Papua New Guinea maintains a special status for anthropological research because of its linguistic and cultural diversity. More than 800 languages are spoken in an area smaller than 463 km^2. The country holds a unique fascination for anthropologists specializing in sex, gender, and sexuality. Anthropologist Bruce Knauft lists more than 20 different researchers who have conducted fieldwork on gender and sexuality in this "object of Western epistemic gaze" (1994: 396).[1] Herdt and Stoller's interest in a Shaman who seemed to them to be intersex, then, is not merely yet another problematic anthropological project located in the anthropologists' mecca of Papa New Guinea. It is emblematic of anthropological research into human sexuality and stands in for the ignorance towards neo_colonial aspects which emerge, when researchers from the Global North infiltrate the Global South to gain knowledge on sensitive issues such as sex-gender-sexuality-systems. What concerns me in this chapter is the nature of Stoller's and Herdt's interaction with Sakulambei, which produces a specific kind of intersexualization—that is cross-cultural intersexualization. Underlying ethnocentric assumptions they carried made them unconsciously treat a Papa New Guinean 'informant/patient' differently than they would a white American subject. Stoller and Herdt, of course, brought to Papua New Guinea a host of assumptions (knowledge) about gender, sex, and sexuality that was generated outside of Papua New Guinea and the Sambian realm.

Looking at an instance of intersexualization in a cross-cultural context lets us compare and contrast it to the kind of intersexualization that happened at John Money's clinic at Johns Hopkins (see Chapter 1 in this volume). Whereas the medicalization and pathologization remain fairly constant, what changes is that the power relations between the participants are entangled in a different context. The power relations between anthropologist/anthropologized, interviewer/interviewee, and psychoanalyst/psychoanalysand prove to be complex and interconnected with mechanisms hard to discern. They can, however, help shed light on the underlying mechanisms in intersexualization—especially in non-colonial settings. By the mid-1980s, Herdt and Stoller had developed their new method of clinical ethnography, yet the overarching structure of anthropological fieldwork—'experts' with authority of the Western university explore/exploit a non-Western culture—remained the same. The crucial means of intersexualization

Anthropology, Sexuality, Neo-colonialism

Early anthropology developed in the cradle of colonialism. Today it is acknowledged that anthropology and ethnology/ethnography descend from a tradition of travel reports and the like. From within anthropology, those self-reflexive about their discipline's complicity with the project of colonialism acknowledge that anthropology carries colonial heritage baggage (e.g., Asad, 1973; Clifford &

Marcus, 1986; Said, 1978). As Talal Asad and the authors of his edited collection *Anthropology and the Colonial Encounter* (1973) first convincingly argued, anthropology can be seen as perpetuating the implicit and explicit power asymmetries of colonialism. Moreover, anthropology arose largely out of the need for colonial administration (Forster, 1973); therefore, a certain synergy between economical interests and anthropology cannot be denied (Grosse, 2000). The origins of anthropology and its methodologies are thus 'colonial,' having emerged in an era when the world was mapped out geographically and politically. This era displayed a particular arrangement of power relations and hierarchies embedded in the expansion of European empires. Colonialism should be regarded as not just a matter of military invasion and economic exploitation; it should also be seen as a practice of imagination through which dominated populations are represented in ways that produce ethnicization/racialization, sexualization, and cultural difference. These modes of knowledge production take place on the level of groups of people and on the level of singular embodiments, subjectivities and practices.

At the most basic level, anthropology seeks to explain cross-cultural variation in human behavior. Anthropologists utilize findings from psychology, medicine, ethnography, archeology, and linguistics to formulate comparative knowledge about humankind. Anthropological interest in sexuality has a long history. When first establishing anthropology as an academic discipline in its own right, early anthropologists, such as Margaret Mead and Bronislaw Malinowski, became aware that sex-gender-sexuality-systems are organizing forms of society. It was not until the 1960s, when taboos regarding sexuality lifted, and studying sexuality became accepted practice. Joseph Carrier, pioneering anthropologist in homosexualities, states in retrospect about the time before the 1960s that "few anthropologists have had the courage to study human sexual behavior in other cultures as a major focus of their research; even fewer have had the courage to study highly stigmatized homosexual behavior" (1986: xi). The American Anthropological Association (AAA) finally acknowledged the importance of sex research at its annual meeting in 1961, when a plenary session was exclusively dedicated to human sexual behavior (see Gebhard, 1971: xiii). In 1974, the AAA sponsored a historic symposium on homosexuality. After this, journals, symposia, and conferences on the issue of (homo)sexuality started to emerge within the field.

Even before that, there were notable exceptions of anthropologists studying sex-gender-sexuality systems. In the early part of the twentieth century, Margaret Mead, along with Bronislaw Malinowski, studied sexuality and gender in a matrilineal island group belonging to Papua New Guinea. Malinowski (1961 [1922]) questioned the idea that Freudian concepts such as the Oedipus complex are universally applicable to all cultures. In questioning the relevance of Oedipus for cultures dissimilar to the nineteenth-century Western European context of Freud, Malinowski implicitly cast doubt on the universal applicability of any knowledge about humans in cross-cultural contexts. Mead's work, also in this island group called the Trobriands, explored whether differences between the sexes_genders were culturally or biologically determined (1930; 1935; 1962 [1950]). Mead's reports on gender-sex-sexuality-systems in Papua New Guinea highly influenced the American sexual revolution of the 1960s, and her early research and ethnographic writings informed later developments in anthropological knowledge of sex, gender, and sexuality.

Inventing Clinical Ethnography

In 1990 Stoller and Herdt published a lengthy description of their professional relationship and their clinical methods over the years. *Intimate Communications* (Herdt & Stoller, 1990) records how they used their "ethnographic dialogues to rethink method" (ix). They first started to rethink how anthropologists compare cultures in the late 1950s, a time period of intense scrutiny in anthropology and ethnography. In the larger academy, poststructural theories had stimulated a general rethinking of method in a number of disciplines. Since the 1950s, poststructuralist theories (namely those by Claude Levi-Strauss) encouraged anthropology's interdisciplinary ties. Anthropologists began incorporating zoology, medicine, biology, cybernetics, pedagogy, history, sociology, and psychology. Herdt and Stoller's contribution to this interdisciplinary agenda was their new method of clinical ethnography.

In their 1987 article, "Der Einfluss der Supervision auf die ethnographische Praxis" (The influence of supervision on ethnographical practice), Herdt and Stoller discuss their introduction of clinical ethnography to the discipline of anthropology. Interestingly, this article in German never appeared in English, however, large parts of it are, however, included in *Intimate Communications* (Herdt & Stoller, 1990). Their new method originated from Stoller's supervision of Herdt's fieldwork from the perspective of a psychiatrist and psychoanalyst. As Herdt learned to overcome his own misogynistic attitude towards the Sambian women by following Stoller's advice to engage with them, they both realized the larger implications this might have. In this 1987 article, Herdt discusses his rising awareness of how he himself influences the set and settings of his fieldwork. However, the article is indeed as it claims rather "ethnography of a supervision" than a reflection on ethnographical practice itself (Herdt & Stoller, 1987: 193). Their conclusion is that the ethnographer should have the same ethical principle of *primum non nocere*—first, do no harm—as physicians do (196). My contention is that they were not able to comply with their own demand in practicing their newly invented method of clinical ethnography.

Under Stoller's influence, Herdt re-conceptualized his own relationship to his informants as a psychoanalyst-psychoanalysand interaction. They introduced Freud's notion of *transference*, the psychoanalytical term for the reproduction of unconscious childhood emotions onto another person (for Freud, the therapist; here, the anthropologist-ethnographer). This enabled them to develop the method of clinical ethnography, albeit only as a halfhearted clinical method. Characteristically, the concept of *countertransference*—where the therapist/ethnographer transfers his own feelings back onto the patient/interviewee—is not mentioned.

By neglecting to account for countertransference, their work assumes influence is a one-way street. And they state that "too much participation has corrupted observation" (51) yet, this only disgraces psychiatry and not anthropology. The Western expert presumably influences his native informant, but not vice versa. The only countertransference taken seriously by Herdt and Stoller is the one of going native (51). The anthropologist, however, is presented as objective, scientific, uninfluenced by his interactions with his informants. Anthropology uses ethnography as its method to compare and analyze the different formations of culture and subjectivity. Many anthropologists re-articulated the aims and methods of anthropology because of the 'epistemological break' that occurred (e.g., Clifford & Marcus, 1986; Rabinow, 1986). Hand in hand with these new impulses to include clinical approaches into ethnography came a new interdisciplinarity. Anthropologists started to collaborate with or receive training by psychologists, psychoanalysts, and also physicians and biologists.

Clinical ethnography, as Stoller and Herdt define it, is a combination of traditional ethnography and psychoanalytical research. Classical ethnography is a method of gathering data about a specific culture and is guided by qualitative and quantitative methods. Clinical ethnography is the concentrated study of subjectivity in cultural context, and focuses on a micro-social understanding of sexual subjectivity and differences in cross-cultural communities. However, Clifford Geertz (1973), in line with other critical anthropologists, questioned the possibility of understanding behaviors and psychological states (conscious and unconscious). Stoller and Herdt, on the contrary, attempted to show that this is still "possible when clinical skill is combined with an in-depth cultural and linguistic understanding of a person and his personality" (Herdt and Stoller 1985: 116). They describe clinical ethnography as a "subtype and more precise form of participant-observation" (116). To them, their "communications with real people, one to one or one to many; people creating and exchanging meanings within interpersonal relationships" is what matters (Herdt & Stoller, 1990: 29). The *clinical* is meant to represent processes of describing, interpreting, and comparing the ways people express feeling, beliefs, and motives. According to Herdt, clinical ethnography is useful for understanding sexual culture because it produces an outcome that tries to understand "the creation of sexual standards as absolutely central to the production of social and personal reality" (Herdt 1999: 110), and it is focused on the "microscopic understanding of sexual subjectivity" (106). With reference to cultural relativism, Herdt asserts that clinical ethnography will "allow for greater diversity in sexuality across cultures, and hence to oppose sexual chauvinism in all its forms" (111). And even though Herdt himself cautions against the clinical diagnosis of 'abnormality' when it comes to sexual cultures (102), his own inclination towards the clinical will nevertheless bring him to exercise the same powerful gesture by adapting the modes of treatment to the informant as 'patient.'

In a footnote, in *Intimate Communications*, Herdt and Stoller note that "the connotation and treatment—of the clinic—is not intended in clinical ethnography," using a dictionary to explain what they mean:

clinical ...involving or depending on direct observation of the living patient; observable by clinical inspection; based on observation, applying objective or standardized methods (as interviews and personality or intelligence tests) to the description, evaluation, and modification of human behavior.

(29)

They add that diagnosis and treatment are not intended in clinical ethnography. Clinical is commonly understood as pertaining to the examination and treatment of a patient. The question I want to pose is, what makes clinical ethnography 'clinical' if the definitions above reflect what ethnography already does—minus the 'patient,' diagnosis, and treatment. Herdt explains that their "work is not therapy (though people felt better for having been able to talk in confidence)" (30). And that clinical ethnography is "like psychoanalysis: best done for years" (31). This seems rather surprising, considering that Stoller stayed for ten days in Papua New Guinea to conduct clinical ethnography with Herdt. The reason I dwell on the 'clinical' in clinical ethnography here is not merely semantic. The 'clinical' in ethnography directly invokes, despite their claims, all of the connotations of the clinic. This includes the implication that a pathological condition exists which is in need of treatment. If we apply this to merely one or two individuals this might not seem so astonishing, but keep in mind this concerns an entire population.

Michel Foucault describes in *The Birth of the Clinic* (1973) how during the nineteenth century medicine, became a disciplining discipline and how biomedicine, the psy-sciences, and related sciences were complicit in creating a control society increasingly characterized by bio-power. The creation of a specific kind of patient—the non-normal—took over and required a new handling; these non-normative people needed to be taken care of by specialists trained by institutionalized knowledge factories, such as the university. Midwives, healers, and charlatans were pushed to the margins; for example knowledge transferred between women for centuries was extinguished. The body/mind split generated a new kind of knowledge in which the only connection between the two was to be expressed solely through psychosomatic pain. The connectedness between body and mind was therefore pathologized whenever it featured. The patient in general became the one who should be lied to about his_her condition because only the specialist knows what the patient 'has.' Particularly in intersex treatment secrecy and silence have been prominent since the 1950s (see Chapter 1 in this volume). Only recently physicians have realized that their 'patients' are very keen to hear the 'truth,' according to which they are treated. The 'clinical' in ethnography, then, pays tribute to the institutionalized, disciplining modes of medicine, and the psy-sciences. Cross-cultural anthropology gained a new feature. A scientific discipline emerged that extends its disciplining powers upon non-Western societies, embodiments, cultures, subjectivities, and bodies. The addition of 'clinical' would make it reasonable to expect to find the data, treatment plan, and cure for an illness. The informant in such an epistemological system is not merely an interviewee but a patient with an illness. Here the patient seems to be an entire culture and not an individual, even though Herdt and Stoller want to make it look like they are 'treating' single individuals. Sakulambei is one of six informants who will be subjected to clinical ethnography by Herdt and Stoller. They use the data however for different means than the reasons given to their interviewees.

Finding the Hermaphrodite Behind the Shaman

Before the interview, Sakulambei is informed that Herdt and Stoller are interested in his shamanic activities and knowledge. The real agenda, however, is to find out about his hermaphroditism. The passivity of Sakulambei that Herdt constructs is immediately striking. Sakulambei's "whatever [Stoller] wants to talk about is all right" is made into a "he'll do whatever you want." Stoller is given the impression that Sakulambei was completely submissive, although the reader can see the discrepancy between what Sakulambei says and what Herdt translates. Herdt seems to try to impress Stoller by stating that Sakulambei will cooperate entirely and that Herdt has been successful making 'friends.' Herdt wants Sakulambei to identify with them so as to extract the secret from him. Herdt wants to learn about [Sakulambei's hermaphroditism/intersexuality], but at first the shaman will not discuss it. Herdt's plan was to use the information "as another kind of 'control' case for understanding the origins and dynamics of Sambia gender identity" (Herdt & Stoller, 1985: 126). The information they intended to extract from Sakulambei was indeed aimed not at understanding intersexuality, it was aimed at understanding 'normal masculinity': "Being a hermaphrodite and so identified in the community, he provided a unique chance for understanding variations and vicissitudes in the development of masculinity" (Herdt & Stoller, 1990: 210). When it proves difficult to obtain the information Herdt and Stoller solicit, they deploy a bit of rhetorical gymnastics to elicit Sakulambei's trust. Herdt wants to know how Sakulambei's sex life is now and complains that he was not able to make him speak about it. Stoller suggests to "do it differently" and to try the following:

S: Tell him that I put my power into you. Would that make sense? Not to the extent that you are me. But that I have put it into you. Not just taught it to you, but [S. makes a noise here like an electric drill—"bzzz": The feeling is implanted in me.] put it into you so that he is safe when he talks to you about these things. [...]

(Herdt & Stoller, 1985: 130)

This excerpt from the transcript is rather frustrating; one cannot really tell which impression Stoller was trying to give. We can only guess what Stoller had in mind when he "bzzzed" Herdt. It seems rather awkward and helpless but Stoller anyway uses this gesture to assure Sakulambei that the information he is going to give them will not be abused. Sakulambei in the situation described has never agreed to subject himself to therapy nor does he know anything beforehand about the situation and what will be asked in the interview. In the relationship between interviewee and interviewer, hierarchies are set in stone. In the *Biographic Portrait* of Sakulambei, Herdt states that "this paper reflects on his [Sakulambei's] life, its public and secret stigmata, especially the private secret he dared not share with anyone, and the relationship with us that allowed him to share" (118). I want to question that Sakulambei was actually willing to share his secret. And I would like to suggest that there have been forces at work that need to be considered in the framework of neo_colonial asymmetries, which are here combined with asymmetries of the newly introduced clinical aspect of ethnography.

"I'm Going to do Something That's Not Quite Ethical"

We can see the dynamic of the catholic confession as explained by Foucault in *The History of Sexuality* (1978). The act of confession purportedly relieves one from the burden of guilt. The confessor is in a position of power over the one who confesses. The whole situation has an absurdity to it—a non-sequitur circle is established in this printed, published, and widely distributed secret of Sakulambei's intersexuality, which Herdt and Stoller promised to keep. Herdt and Stoller seem ignorant to this fact of their breaking the promise as they exploit and misuse Sakulambei's trust for the sake of their research. The intimate surveillance of Sakulambei constitutes his sex as a secret that has to be discovered. To paraphrase Armstrong, Herdt and Stoller finely tune "the surveillance machinery" (1987: 72); with their interview, Herdt and Stoller seek to capture the essence of the confession. The intimate surveillance, which Herdt and Stoller conduct over Sakulambei constitutes his sex as a secret that has to be discovered by them. This puts the researchers in position of knowing, though this position of knowing is much more powerful than that of the interviewee: he is not informed about their aspirations. Sakulambei presumably had nothing to confess before Herdt and Stoller arrived and made him do so. The interview is transcribed in detail and contains a number of sequences where Stoller and Herdt speculate how Sakulambei might feel and if he wants permission by Stoller to cry. When Herdt states that crying is not acceptable in Sakulambei's culture, Stoller says that it would be alright with him if Sakulambei cried (Herdt & Stoller, 1985: 147).

The colonizing paternalism displayed here is typical in Herdt's and Stoller's ethnographies. Representing Sakulambei as childlike denies him full subject status. Herdt states that "Saku's sense of self is so steeped in his shamanic role that his selfhood is merged with that role, and I believe he is unconscious of how much he defends himself and denies his past through this identity" (Herdt & Stoller, 1990: 373). However, Sakulambei is anything but unconscious about defending himself against the intrusive questions. Herdt and Stoller posit an a priori difference by exoticizing and 'making the native speak,' whereas they simultaneously create a silence. Sakulambei cannot be heard through Herdt's and Stoller's constant talk and interpretation. Stoller guides Herdt through the process of questioning the interviewee by telling him to assure him that they are not going to harm him.

According to Herdt, a masculine man has a penis and erections, and he uses this penis to engage in reproductive sex with women (134). The biographical note about Sakulambei observes that "Saku has always been known as an enthusiastic and skillful fellator" (in initiation rites) and that he continued to be a fellator after the normal period for initiates. This makes his masculinity suspect for Herdt (123). This was also the initial reason for Herdt to suspect Sakulambei's intersexuality. Sakulambei's not following the masculine protocol of shifting from fellating to being fellated implies femininity—or at least an incomplete masculinity. The assumed nexus between a person's sexual orientation/preference, and his_her body and behavior serves here as a central point of reference, which mirrors the categories that Herdt and Stoller use to conceptualize Sakulambei as a male hermaphrodite.

In order to make Sakulambei speak of his presumed hermaphroditism, Stoller and Herdt adapt rather unethical methods of letting Sakulambei think that first, in their culture, only special people become shamans and second that because Stoller is also a hermaphrodite he is also a shaman. The secret he ultimately reveals is elicited through a lie; yet they admit that it is quite unethical—they however chose to let Sakulambei believe that Stoller also has "some hermaphroditic qualities." Stoller even admits that he has done this before while working with patients, which is why he felt that it was not "unethical" (134). He admits to have done exactly the same to his patients at home—leading them to or letting them think that he is also intersex. Herdt therefore implies that Stoller is also a hermaphrodite and that, therefore, he can trust him. Sakulambei is invited to identify with Stoller, and thus form an imagined community with Stoller. This seems to be a rather unethical practice when it comes to psychoanalysis and psychotherapy. Stoller, however, does not question telling a fundamental lie to his patients—neither in the US nor in Papua New Guinea—in order to make them trust him. Despite acknowledging their own questionable ethics, the two researchers continue and shift the responsibility for ethical conduct as a researcher at least in being honest to the informant's ability to project this narrative of being intersex onto himself:

H: [To Sa.] His appearance is real, but when he [S.] was born, they looked at this body and wondered: "This baby is another kind," or... They didn't know for sure. Now, when you see him, you see that he's become the same as a man. But now he's become a shaman. And now he's a man, too. He's the same as you... but when he was born, they thought he was a different kind. And so he wants to know if it was the same with you.

SA: This. ... I don't really understand you. [Pause. Silence.] My *kwooluku* ... when, before my mother gave birth. I don't understand well. [He understands precisely but is dodging. Still, compared to Saku's earlier sessions with H., he is not now frantic. Rather his voice is calm; he is not frightened (comment by Herdt)].

(Herdt & Stoller, 1985: 135)

Unfortunately these earlier conversations where Sakulambei was frantic are not included. However, Herdt thinks that Sakulambei is evading their questions. As I read it, Sakulambei seems to have thought that they wanted to know more about how he became a shaman because he constantly refers to his profession as a shaman; this is understandable, since this is what he was told the interview was about. Larger parts of the interview in which Sakulambei talks about his shamanism have obviously been removed from the transcript, since for Herdt and Stoller these descriptions lead them from the main course of finding out about his hermaphroditism/intersexuality.

Sex is given prominence as the secret of the Western gaze. This Western gaze is in the case of clinical ethnography also a clinical gaze as described by Foucault (1973). The pathology of Sakulambei's sex is what Herdt and Stoller are after and what makes them blind towards other information that he might want to give. The "different kind" that is evoked here by Herdt might not be related to sex at all for Sakulambei. Sakulambei's rhetorical strategies (i.e., dodging) might be a sign that he realizes what they are getting at. The Foucauldian 'truth' of Sakulambei's body is at stake here. Can we assume that Herdt's expectation of what he was to get out of Sakulambei guided the mode of interrogation? Sakulambei hardly says anything. It seems as if Sakulambei capitulates to their view in order to extract himself from the awkward situation. Sakulambei eventually tells them:

SA: They all looked at us at first—"I think it's a girl," that's what they thought. And then, later, they all looked at us and saw that we had a ball [testes], and they all said: I think it's a male.[...] And, likewise we've got cocks. .. and we've got balls. [...] But our water [urine] we all lose it in the middle [extreme hypospadias: urinary meatus in female positions, not at the distal end of glans penis as in normal males] (*sic!*).

(Herdt & Stoller, 1985: 140–141)

The researchers expect Sakulambei to describe his body in detail but their hopes are disappointed. They form the medical diagnosis of extreme hypospadias according only to this limited confession. Sakulambei has finally confessed to having an unruly body, which now can be designated to be intersex. Herdt and Stoller do not refrain from medicalizing and pathologizing Sakulambei's body according to the medical standards of the West.

The Other is Still a Fiction of the Ethnographer's Own Making

One key aspect of male personhood amongst the Sambia is the participation in initiation rituals. Herdt observed that the Sambia only rarely initiate *kwolu-aatmwol* (the indigenous term for intersex*) into third-stage *ipmangwi* (puberty) status.[2] However, at one point Herdt states that he witnessed Sakulambei's initiation into the puberty stage in the 1970s. Yet, Sakulambei's own cohort does not do this, which for Herdt is a sign that Sakulambei's history is supporting "the interpretation of a three-sex code in subsequent socialization" (Herdt, 1994: 439). Herdt believes that the reason for this is because "biological changes in the male body anticipate the subsequent social events" (439) of this third-stage rite which cannot provide the *kwolu-aatmwol* with the power to complete masculinization. Sakulambei has never taken part in the nose-bleeding ritual, which is believed to drain an overabundance of female blood from the body. Herdt suggests that Sakulambei feels that this female element is "part of his core identity" (439). He concludes that:

> it is no surprise then, that the *kwolu-aatmwol's* gender-identity state is neither clearly male nor female: they have a hermaphroditic psychosexual identity that is distinctly different, and their phenomenology reveals them to feel unique or alone in the world.
>
> (440)

The concept of a 'psychosexual identity' invokes Freudian psychoanalysis and therefore specific Western discourses on the self. Herdt's use of the terms 'core identity' and 'hermaphroditic psychosexual identity' reflect Stoller's influence on his thinking.

Sakulambei uses the words WE/US when he talks about the process of determining his gender when he was born. Herdt and Stoller do not elaborate on this rather interesting fact that Sakulambei switches to the plural when he talks about his body. I do not want to give an interpretation of this, though I suggest that the use of the plural might denote a different conceptualization of the relation of the self to the body and might stress a different version of I and embodiment than the one of the West. This passage might suggest that in the Sambian understanding of the individual/person/I/self/embodiment it is not possible to separate the person from his_her environment (see also Chapter 5 in this volume).

The conception of the self in the Western sense displays a certain focus, which has been explained by Foucault in the *History of Sexuality* (1978), in which he argues that sex is constructed as the secret of the self that can be discovered anywhere. A precondition to this conceptualization and perception of the self is a self that is I—an 'individual.' Especially self-other relations are constructed differently in most cultures but also the construction of the self, which is intrinsically connected to the semiotics that are available in specific languages. Feminist anthropologist Gloria Wekker elaborates on the different ways of being in the world that different languages cause; her example is a rather unexpected one and illustrates her point. According to her, "the speakers of American and British English, however close those languages varieties may be, inhabit different worlds" (Wekker & Wekker, 1991). From this critical perspective:

> Western thinking about personhood is, postmodernity notwithstanding, still characterized by two important commonsense assumptions, which also inform academic studies. In the first place, there is the assumption that a "normal," healthy individual has a fixed core to herself or, more typically, to

himself. Even though the situation in which this individual finds himself or herself may change, he or she remains the same person. Individuals, in other words, are seen as fixed, static. The second assumption about the self is that it is bounded, that there is a clear and "natural" distinction and juxtaposition between the individual and society. The preoccupation with an essential interior "I" keeps the boundaries between this inner "I" and outside world intact. Thinking in terms of these two assumptions is not only deeply rooted, it is still often projected unto selfhood in other societies. The person-as-an-island approach constantly reproduces notions of uniqueness, unchange-ability, and boundedness.

(Wekker, 2006: 102/103)

Therefore, the problem of translation of the self and conception of subjectivity has to be radicalized when it comes to the attempt to translate psychic structures from one society to another. Neither psychoanalytical nor bio-medical explana-tions of the West make sense when it comes to the embodiment of a person whose cultural background and semiotics do not provide a concept of the self that is an individual, who has one singular body and identity.

Neo_colonial interpretations of the 'other' do not necessarily have to be geo-graphical. They are, however, always political. The mechanisms at work in the production of knowledge on 'other' cultures have been conceptualized as Euro-centrism, Orientalism, or Occidentalism (e.g., Said, 1978; Venn, 1999; Young, 1990). These notions designate the creation of two distinct epistemological spaces: the 'West and the Rest' (Hall, 1996). This division of space is no longer and has never truly been a geographical one. Specific effects of hegemonic rep-resentations of the Western self and culture produce the other as distinct in a specific way. Therefore, the processes of institutionalization and legitimization of anthropological knowledge can be seen in light of Foucault's power/knowledge complex. Everything vague is measured against the parameter of a constant affir-mation of the 'West.' Elspeth Probyn has examined "the position of the knower in relation to the stories that he tells" (1993: 61). She argues that the anthropo-logical other "is but an effect of the ontological construction of the ethnogra-pher's self; the other is still a fiction of the ethnographer's own making" (66). Stoller and Herdt project the phantasm of hermaphroditism/intersexuality upon Sakulambei. When they describe Sakulambei's body as showing extreme hypos-padias their power to define is a self-affirmation in regards to their expectations of what Sakulambei will reveal to them. By the end of the interview, we know more about the 'knowers' than we know about Sakulambei.

Look, Listen, and Say

Toward the end of the interview, we learn that a German businessman named Gronemann arrived at the village about 20 years earlier, when Sakulambei was a child. Gronemann came with a government control agency passing through the Sambia valley in the 1960s. He undressed the 10-year-old Sakulambei and took pictures of him. For Sakulambei this has happened without reason. He also states that he was not paid for allowing the picture to be taken (Herdt and Stoller 1985: 144). Stoller realizes that he has been behaving in much the same manner. He assures the reader of the following: "I will *not* take a photograph; I will not ever humiliate him [...]" (Herdt & Stoller, 1985: 145).

The structure of Herdt and Stoller's encounter with Sakulambei is not so different from the ten-year-old's encounter with the German. Both interactions take place unwillingly, and Sakulambei is in both cases unpaid. In the case of Herdt and Stoller we can say that Sakulambei never really wanted to talk about his body—he wanted to talk about his profession as a shaman and was more or less tricked into talking about his intersexuality. The information he was however willing to convey—the one about his shamanism—has been deleted from the published account.

Gronemann's publication includes a naked photograph with the patient's eyes blacked out to conceal identity. Intersex activist Cheryl Chase has been quoted by Alice Dreger to say that "the only thing the black band over the eyes accomplishes is saving the viewer from having the subject stare back" (Dreger, 2000: 162). Dreger published *Intersex in the Age of Ethics* (1999) with a picture of herself—a non-intersexualized person—on the cover, with eyes blacked out. She did this in order to show that everyone could look "rather pathological if photographed in this way" (2000: 162). The mode of displaying a body influences not just the one who views it but also the one who is shown by it. And it is of course not just the specificity of the image with the blacked out eyes that produces the air of pathology and accompanying shame. The experience of being examined and diagnosed induces feelings of shame and humiliation (Creighton, Alderson, Minto, & Brown, 2002; Dreger 1999). Sakulambei's experience with Gronemann has clearly evoked similar emotions. If asked, what would he say about the image of him that was distributed in anthropological publications and only obscured by a pseudonym? In both the word and image, ethnographers rely heavily on the objectifying 'technologies of realism' (Ball & Smith, 2001). As much as positivism and photography go hand in hand, bio-medical and psychological data tend to be similarly positivistic. According to Ann McClintock a feminist scholar "photography was both a technology of representation and a technology of power" (1995: 124). With photography, "Western knowledge and Western authority became synonymous with the real" (123). McClintock reminds us of the role that photography played in the context of imperialism:

> it should not be forgotten that photography emerged as a technology of surveillance within the context of a developing global economy. A circulation of notions can be observed between photography and imperialism.(...) Photography provided the cultural equivalent of a universal currency. (...) Hailed as superseding the messy enigmas of language and as capable of communication on a global scale through the universal faculty of vision, photography shifted the authority of universal knowledge from print language to spectacle.

(123)

The realistic impetus of both clinical data and pictorial methods produces notions of objectivity and truth of ethnographic data and "thus capitalizes upon [its] immense descriptive potential" (Ball & Smith, 2001: 304). Ball & Smith furthermore describe the realism of documentary, along which I also count the clinical data, as a "professional ideology" (304). As much as photography or film has been used in anthropology to establish certain assumptions, clinical data serve the manifestation of specific fantasies. Ball & Smith state:

> In the late nineteenth century, influenced by pre-Darwinian evolutionist theories, physical anthropology and anthropometry made extensive use of photography to reveal the putative difference between the Mongol, Negro and Caucasian "racial" groups. Guided by Huxley and Lamprey's attempts to systematize and record the physiological measurement of body mass and skeletal size in a manner that would enable reliable comparative morphometric data to be collected, anthropometric photography became established.
>
> (2001: 306)

The 'myth of photographic truth' (Sekula, 1975), which emerges by presenting photography as unmediated, mechanical transcriptions of transparent reality is also apparent in the clinical data presented by Herdt and Stoller. The photograph invokes realism; it emphasizes the affinity between physiognomy and a supposedly universal language (Fyfe & Law, 1988). Photographic data derive their power and autonomy from the discourses of bio-power and posit an assumed physical difference at the core of ethnographic fieldwork.[3] Herdt and Stoller do not reprint the photographs. They refer to Groneman's pictures or rather the invocation of the picture. Even though they do not reprint them, they do install them as a powerful reference by referring to them as the proof for Sakulambei's hermaphroditism. The power of the picture which was taken 30 years ago is reinstalled solely by the reference and the reliance that Gronemann's gaze can be trusted, i.e. is real(istic). The connection between photographic pictures and medical clinical data, and diagnosis is that both methods (or concepts) appear to provide the most objective and truthful account of the real. The authenticity of the picture is invoked, medical diagnosis is added, and we are seduced to believe.

In *The Birth of the Clinic* Foucault describes the workings of the gaze in relation to hearing and states that:

> a hearing gaze and a seeing gaze: clinical experience represents a moment of balance between speech and spectacle. A precarious balance, for it rests on a formidable postulate: that all that is *visible* is *expressible*, and that it is *wholly visible* because it is *wholly expressible*.
>
> (1973: 115 [emphasis in original])

Foucault describes here the power of the photograph in relation to the verbal description of the 'fact' being visually portrayed. The interlinking of both, the visual and the verbal portraying the same issue, produces even more power; vision and hearing interlocking generate a powerful truth claim supporting scientific statements and their knowledge production. Herdt and Stoller's interpretation of Sakulambei's body can be understood along these lines. The medical diagnosis that they want to 'hear' in the interview interlinks with the visible 'spectacle' and forms the basis for the diagnosis to be made. Sakulambei's spoken account is for them proof for the materiality of the body. The historical photograph that has been taken by Gronemann—the fact that the picture has been taken at all—already testifies to the existence of something that needs to be displayed as exceptional. Vision is firmly intertwined with language. Anthropology combined with clinical ethnography posits severe challenges to translation and produces pathologization of interviewees/informants/patients—either as individuals or as a group.

For feminist scholars Sandra Harding and Merril Hintikka, positivism is based on the 'seeing eye' (Harding & Hintikka, 1983). The 'seeing eye' is the tool for the observation by the traveler/intruder/anthropologist and the observation serves as the foundation of distinctions and those serve the process of classification. Visible nature is to be distinguished and classified according to the gaze that the travelers/intruders/anthropologists bring with them. According to Foucault this is the ordering of the science which happens through "the exercise and decision" of this gaze (1973: 89). He has described this gaze as "the gaze that sees" and which inevitably "is a gaze that dominates" (39). The gaze that Foucault describes is not the poetic and benevolent gaze of longing. Rather, he describes the gaze of psycho-medical discourse; a purposeful looking that creates meanings and their boundaries more than it just sees them. In Western tradition visual perception is considered to be the privileged means of access to 'reality'—to see is to know. Especially photography (or in this case, the mere mention of it) is assumed to invoke a 'real' referent. Donna Haraway has created the notion of the 'camera-gun' to depict the colonialist violence exercised by photographic imagery (1989). "Ultimately," she states, is the camera "so superior to the gun for the possession, production, reservation, consumption, surveillance, appreciation, and control of nature" (1989: 45). The perfect companion for anthropological science is therefore the photographic picture. The psycho-medical and anthropological narratives I analyzed are, as Valerie Traub describes it, "both dedicated to rendering intelligible and distinct that which appears chaotic, primitive, or previously unknown, through strategies of description, nomination, and classification" (1999: 305). When it comes to gazing—either as clinical, Western, hearing, or seeing: "to gaze implies more than to look at—it signifies a psychological relationship of power, in which the gazer is superior to the object of the gaze" (Schroeder, 1998: 208). The question is always who is entitled to gaze and who is looked or stared at. The clinical ethnographic account by Herdt and Stoller penetrates Sakulambei's body especially by the gaze. Their clinical gaze infiltrates Sakulambei in combination with their interrogational mode, their intrusive examination and psychoanalytical interrogation. Herdt and Stoller undertake all these measures in order to find their suspicion of intersex confirmed—intersexualization completed.

Traub furthermore argues that "locating the body (and bodies) within prevailing epistemic hierarchies by charting corporeal cartographies, anatomies and travel narratives not only function as colonialist discourses but urge colonialism into being" (Traub, 1999: 305). When intertwined with sexuality these processes are laden with relations of power. Sex-gender-sexuality systems are major avenues through which so-called experts exercise control over patients. Anthropologists Lyons and Lyons further suggest that in "Foucault's formulation the watchers do not create sexual behaviors or insane ideation. They label them, diagnose them, and give social reality" (2004: 13). According to them, sexuality can secure a "regime of bodily control, categorizing and disciplining behavior and identity" (12–13). In terms of intersexualization the watchers or researchers are complicit in a creation process, not least regarding terminology. Viola Amato has recently argued that in intersex research "visual presentation, the physical appearance of specific bodily characteristic, becomes the basis for cultural claims, i.e. how to classify the body according to a normative, binary gender system" (2016: 46) and that "heightened visibility of intersex bodies (…) entails their invisibilization" (49). One important accompanying feature to the invisibilization of Sakulambei that happens in the interview is the silencing of Sakulambei.

Anthropology in Translation

By referring to Edward Evans-Pritchard (1962) and Clifford Geertz (1973), Herdt and Stoller state that anthropology is translation, adding that "translation is evocation only; translations are interpretations, editings" (Herdt & Stoller, 1990: 88). Stoller neither speaks the language of the Sambia nor Pidgin, so he cannot communicate directly with his interviewee. Herdt translates (partially and rephrased) Pidgin. He presents Stoller as an American 'shaman' to the Sambian shaman Sakulambei in the hope he will identify with Stoller, or at least perceive Stoller as a colleague he can trust. Pidgin is Sakulambei's second language, so Stoller and Herdt both face major interpretive obstacles in transmitting information. In his review of *Intimate Communications*, anthropologist Alfred Gell writes that that the "level of personal or cultural insight communicated in Herdt's interviews is low indeed. And what any of this," Gell wonders, "has to do with properly conducted psychotherapy beats me. A common language would appear to be prerequisite; but not, evidently for 'clinical anthropology'" (Gell, 1993: 839). They are ignorant towards the semantic differences between their own culture's language, Pidgin and the Sambian language. Translation however, is always enacting a kind of 'violence' on language. It is always only an approximation at best.

Surprisingly Herdt and Stoller, even though they try to consider many aspects of the combination of the clinical with ethnography, do not explicitly reflect on the "ethnocentric violence of translation" (Venuti, 1995: 810), nor do they incorporate the question of power in the neo_colonial setting they are engaged in. Various aspects of hierarchy, hegemony, and cultural dominance are reflected in neo_colonial settings—translation seems to be intimately bound up with questions of power, politics, and identity. The issue of translation has long been extensively discussed in anthropology. Boas (1921) already recognized that the organization of languages of American Native tribes of the New World were different than European languages and Latin. Malinowski, in his Introduction to *Argonauts of the Western Pacific* (1961 [1922]) noted that 'pidgin English' was very limited in its usefulness in acquiring knowledge. In their Introduction to *Translating Cultures* (2003) Paula Rubel and Abraham Rosman state that translation is central to "writing about culture" (1), since translation means cross-cultural understanding. They declare furthermore that "translation is doomed to inadequacy because of irreducible differences not only between languages and cultures, but within them as well" (10). However, they do confirm that "a form of translation can still take place" (11). From the ethnographer, they demand that the inadequacies of the translation should be met with great caution and recommend that commentaries with reflections on translation issues accompany the ethnography. They ask for reflections on the transformations that take place in translations on the semantic, syntactic, and discursive levels. James Clifford also noted that the ethnographer should have an appreciation of what is missed in any translation and what is necessarily deemed to be inaccurate in the very act of understanding and describing another culture (1997). There is also another kind of translation, which ethnographers perform than the 'purely' linguistic one. Ethnographers will always 'translate' what they found on the local level into analytical concepts. In order to compare societies, their similarities and differences, the ethnographer will develop cultural categories such as kinship, social organization, warfare, nutrition, rituals, etc. These will necessarily be limited.

Coming back to Herdt and Stoller, the translation of affect and psychological states is yet another issue which becomes salient. Any translator crosses and thereby violates boundaries and the intimacy of the cultural setting within which he or she is working. Yet, as Michael Herzfeld (2003) argues, although translation crosses boundaries it also creates boundaries and it could also be seen as a betrayal since it violates the cultural intimacy of the 'inside.' This 'inside' is of course present on a variety of levels: on the level of the individual, the direct group-context of the individual, the wider group, etc. However, Herdt and Stoller do not consider issues of translation, and they also do not reflect on Sakulambei's intimacy and the power as well as the violation they might exert in conducting the interview.

A translation from Pidgin to English by Herdt of the conversation between Stoller and Sakulambei was necessary since Stoller did not speak Sakulambei's language. Herdt and Stoller have discussions in English in front of Sakulambei, and Herdt decides what to translate for Sakulambei. Herdt himself speaks Pidgin, which is Sakulambei's second language, so Stoller and Herdt both face major interpretive obstacles in transmitting information. Alfred Gell reasons concerning this fact, in his review of *Intimate Communications*, that the "level of personal or cultural insight communicated in Herdt's interviews is low indeed. And what any of this," Gell wonders, "has to do with properly conducted psychotherapy beats me. A common language would appear to be prerequisite; but not, evidently for 'clinical anthropology'" (1993: 839). Indeed it is especially striking that no translation or transparency is required; a direct transferability of Sakulambei's psychic structures is assumed, which means here that psychic structures and their explanatory models are presumed to be cross-culturally valid and therefore essential. This alleged essential feature of psychic models is exercised on the basis of a cultural difference, which in its social implementation serves as the precondition for the exercise for clinical ethnography. This assumption reinforces the supposed universality of psychic structures and their pathologies because Herdt and Stoller assume transferability.

Conclusion

Herdt's and Stoller's newly invented method of clinical ethnography—developed in part through their study of Sakulambei—is significant in intersex history because it takes intersexualization into a cross-cultural i.e., neo_colonial context. I focus on the interview with Sakulambei at length because it illuminates specific aspects of the processes of medicalization and pathologization in intersexualizations. In intersexualization the visual is central in constructing bodies as non-normal and modes of gazing at bodies exert the power to diagnose. In cross-cultural intersexualization it is not just the clinical gaze but also the hegemonic Western gaze that is employed to dissect and categorize the body of the 'other.' The normalizing powers of the clinical gaze in clinical ethnography are multiplied because local meanings of bodies and psyches are overridden by neo_colonial modes of knowledge production. By bringing psychoanalytic methods to an ethnographic study of a Sambian hermaphrodite, Herdt and Stoller transfer the power of the clinical into a realm that has formerly not been subjected to the normalizing forces of Western medicine and its control mechanisms. With their new method, Herdt and Stoller collapse and conflate the position of the anthropological informant with that of the psychoanalysand. This conflation sheds light on the clinical ethnographers rather than the culture/individual under investigation. Moreover, the initial lie they tell Sakulambei about Stoller's non-existent 'intersex condition' to make Sakulambei talk is particularly significant, not just because it shows them to be unethical, but because it proves them to be paternalistic. Sakulambei's body has been colonized by Western modes of research and Western assumptions about his bodily features and alleged psychological 'condition,' which results in cross-cultural intersexualization. Sakulambei is silenced the very minute he is translated to English. Even though Herdt and Stoller acknowledge that anthropology is translation they do not pay justice to this fact. They neither problematize that they do not speak the same language as Sakulambei, nor that the limited means they use for communication—Pidgin English—may pose serious limitations to their clinical ethnography. They also do not contemplate the sensitive issue of talking about a secret that is connected with shame and the experience of having been subjected to a similar situation earlier in his life. Sakulambei's photo has been taken when he was a child; he has therefore been singled out. His word on the one hand does count, and on the other is disregarded, because the fact that the picture has been taken is used to solidify

Herdt's and Stoller's claim, that he is intersex*. Sakulambei is subjected to the Western processes of (inter)sexualization through this twofold process of silencing and invisibilizing and the accompanying modes of pathologization.

The case of Caster Semenya that I opened this book with shows a contemporary instance of intersexualization. In her case, a twenty-first century competitive athlete is silenced and pathologized because her voice cannot be heard in the multiple media coverage. Semenya's body has also at the same time been excessively displayed and viewed. She is also forced to confess to the truth of her sex—or at least convinced to subject and comply to Western standards of femininity—as on the cover of *YOU*—with painted fingernails, lipstick, and long hair. Sakulambei's confession of his bodily features is consolidated by his confession of having been photographed naked by another Westerner. Like the sports committee for Semenya, Herdt and Stoller have taken over the task of designating Sakulambei's fate as either a man, a woman, or an intersexualized person. Unlike Semenya's case, Sakulambei's shows how the interviewee has been encouraged to talk about and identify with a specific body, a specific sex_gender and a specific sex(uality), which Herdt and Stoller have laid out for him.

One difference between intersex and intersexualization, as I have already demonstrated in Chapters 1 and 2 in this volume, is what is considered to be the phenomenon. In intersex it is the intersex* body that helps distinguish between the categories of male and female, and in intersexualization it is the researcher that is considered to be the phenomenon. The term intersex suggests that the body in question is non-normative or even non-natural whereas intersexualization draws attention to the fact that the phenomenon is the experts' desire to pathologize and medicalize. In cross-cultural intersexualization we are additionally faced with the desire of the expert to universalize Western systems of sex, gender, sexuality, and notions of personhood.

Notes

1. Bruce Knauft described Melanesia as a vibrantly changing world area, and anthropology as a vibrantly changing discipline, which makes the relationship between the two highly complex (1999: 3). In an issue of *Current Anthropology* of 1993, Terrence Hays and his respondents reflect on the distinctions between the High Lands and the Low Lands of Papua New Guinea, and controversially discuss and categorize Papua New Guinea as a 'fuzzy set.' Some anthropologists stress that these determinations imply homogeneity and are therefore misleading in terms of essentialist notions of a highly diverse region Hays, Brown, Harrison, Hauser-Schäublin, & Hayano, et al. 1993). Since the first colonialist arrived in Papua New Guinea, the resulting ethnography on this region tended to suspend subjectivity and replace it "by a plethora of particular, separate, and impersonal beliefs and customs" (Knauft, 1994: 400).
2. Herdt has published widely on the rites of passage in Sambian culture (e.g., Herdt, 1981, 1984).
3. In cross-cultural anthropology the use of photography also represents another dimension, which I will not discuss here, due to space. In regards to racialization and the use of photography in anthropology, Deborah Poole states that "photographs were rendered as 'data'" and through this "the concept of race emerged as an abstraction produced by the archive as a technological form" (Poole, 2005).

Bibliography

Amato, V. (2016). *Intersex narratives shifts in the representation of intersex lives in North American literature and popular Culture*. Bielefeld: transcript Verlag.

Armstrong, D. (1987). Bodies of knowledge: Foucault and the problem of human anatomy. In G. Scrambler (Ed.), *Sociological theory and medical sociology* (pp. 59–76). London, New York: Travistock Publications.

Asad, T. (ed.) (1973). *Anthropology and the colonial encounter*. London: Ithaca, 23–38.

Ball, M., & Smith, G. (2001). Technologies of realism? Ethnographic uses of photography and film. In P. Atkinson, A. Coffey, S. Delamont, & J. Lofland (Eds.), *Handbook of Ethnography* (pp. 302–319). London: Sage.

Boas, F. (1921). *Ethnology of the Kwakiutl*. In Thirty-fifth annual report oft the Bureau of American Ethnology to the secretary of the Smithsonian Institution 1913–1914 (pp. 41–794). Washington: Government Printing Office.

Carrier, J. M. (1986). Foreword. In E. Blackwood (Ed.), *The many faces of homosexuality. Anthropological approaches to homosexual behaviour* (pp. xi–xii). New York, London: Harrington Park Press.

Clifford, J. (1997). *Routes: Travel and translation in the late twentieth century*. Cambridge: Harvard University Press.

Clifford, J., & Marcus, G. (Eds.). (1986). *Writing culture: The poetics and politics of ethnography*. Berkeley & London: University of California Press.

Creighton, S., Alderson, J., Minto, C. L., & Brown, S. (2002). Medical photography: Ethics, consent and the intersex patient. *BJU International, 89*(1), 67–72.

Dreger, A. D. (Ed.). (1999). *Intersex in the age of ethics*. Hagerstown, Maryland: University Publishing Group.

Dreger, A. (2000). Jarring bodies: Thoughts on the display of unusual anatomies. *Perspectives in biology and medicine, 43*(2), 161–172.

Evans-Pritchard, E. E. (1962). *Social anthropology and other essays: Combining social anthropology and essays in social anthropology*. New York: Free Press of Glencoe.

Forster, P. (1973). Empiricism and imperialism. In T. Asad (Ed.), *Anthropology and the colonial encounter* (pp. 23–38). London: Ithaca.

Foucault, M. (1973). *The birth of the clinic. An archaeology of medical perception*. London: Travistock Publications.

Foucault, M. (1978). *History of sexuality*. London: Penguin Books.

Fyfe, G., & Law, J. (Eds.). (1988). *Picturing power: Visual depictions and social relations*. New York: Routledge.

Gebhard, P. (1971). Foreword. In D. Marshall, & R. Suggs (Eds.), *Human sexual behavior: variations in the ethnographic spectrum* (pp. xi–xiv). New York: Basic Books.

Geertz, C. (1973). *The interpretation of cultures*. New York: Basic Books.

Gell, A. F. (1993). Review: Intimate communications: Erotics and the study of culture by Gilbert Herdt and Robert Stoller. *Man, 28*(4), 838–840.

Grosse, P. (2000). *Kolonialismus, Eugenik und bürgerliche Gesellschaft in Deutschland 1850–1918*. Frankfurt am Main/New York: Campus.

Hall, S. (1996). The west and the rest: Discourse and power. In S. Hall, & B. Grieben (Eds.), *Modernity: An introduction to modern societies* (pp. 275–320). Oxford: Wiley-Blackwell.

Haraway, D. (1989). *Primate visions: Gender, race, and nature in the world of modern science*. New York: Routledge.

Harding, S., & Hintikka, M. B. (Eds.). (1983). *Discovering reality. Feminist perspectives on epistemology, metaphysics, methodology, and philosophy of science.* Dordrecht, Boston, London: D. Reidel Publishing Company.

Hays, T., Brown, P., Harrison, G. W. S., Hauser-Schäublin, B., Hayano, D., Hirsch, E., et al. (1993). "The New Guinea Highlands": Region, culture area, or fuzzy set? [and comments and reply]. In *Current Anthropology, 34*(2), 141–164.

Herdt, G. (1981). *Guardians of the flutes: Idioms of masculinity.* New York: McGraw-Hill.

Herdt, G. (1982). *Rituals of manhood.* Berkeley: University of California Press.

Herdt, G. (Ed.). (1984). *Ritualized homosexuality in Melanesia.* Berkeley: University of California Press.

Herdt, G. (1994). Mistaken sex: Culture, biology and the third sex in New Guinea. In G. Herdt (Ed.), *Third sex third gender. Beyond sexual dimorphism in culture and history* (pp. 419–446). New York: Zone Books.

Herdt, G. (1999). Clinical ethnography and sexual culture. *Annual Review of Sex Research, 10,* 100–119.

Herdt, G, & Stoller, R. (1985). Sakulambei—A hermaphrodite's secret: An example of clinical ethnography. *Psychoanalytic Study of Society, 11,* 117–158.

Herdt, G., & Stoller, R. (1987). Der Einfluss der Supervision auf die ethnographische Praxis. In H. P. Duerr (Ed.), *Die wilde Seele* (pp. 177–199). Frankfurt am Main: Suhrkamp.

Herdt, G., & Stoller, R. (1990). *Intimate communications: Erotics and the study of culture.* New York: Columbia University Press.

Herzfeld, M. (2003). The unspeakable in pursuit of the ineffable: Representations of untranslatability in ethnographic discourse. In P. G. Rubel, & A. Rosman (Eds.), *Translating cultures perspectives on translation and anthropology* (pp. 109–134). Oxford, New York: Berg.

Knauft, B. M. (1994). Foucault meets South New Guinea: Knowledge, power, sexuality. *Ethos, 22*(4), 391–438.

Knauft, B. M. (1999). *From primitive to postcolonial in Melanesia and anthropology.* Ann Arbor: University of Michigan Press.

Lyons, A., & Lyons, H. (2004). *Irregular connections. A history of anthropology and sexuality.* Lincoln, London: University of Nebraska Press.

Malinowski, B. (1961 [1922]). *Argonauts of the Western Pacific.* New York: E. P. Dutton.

McClintock, A. (1995). *Imperial leather. Race, gender and sexuality in the colonial contest.* New York, London: Routledge.

Mead, M. (1930). *Growing up in New Guinea.* New York: Blue Ribbon.

Mead, M. (1935). *Sex and temperament in three primitive societies.* New York: William Morrow.

Mead, M. (1962 [1950]). *Male and female. A study of the sexes in a changing world.* Harmondsworth: Penguin Books.

Poole, D. (2005). An excess of description: Ethnography, race, and visual technologies. *Annual Review of Anthropology 10*(34), 159–179.

Probyn, E. (1993). *Sexing the self.* London: Routledge.

Rabinow, P. (1986). Representations Are social facts: Modernity and post-modernity in anthropology. In J. Clifford and G. Marcus (Eds.), *Writing culture. The poetics and politics of ethnography* (pp. 234–261). Berkeley, London: University of California Press.

Rubel, P. G., & Rosman, A. (Eds.). (2003). *Translating cultures perspectives on translation and anthropology.* Oxford, New York: Berg.

Said, E. (1978). *Orientalism*. New York: Vintage Books.

Schroeder, J. (1998). Consuming representation: A visual approach to consumer research. In Barbara B. Stern (Ed.), *Representing consumers: Voices, views and visions* (pp. 193–230). London: Routledge.

Sekula, A. (1975). On the invention of photographic meaning. In *Artforum, 13*(5), 36–45.

Traub, V. (1999). The psychomorphology of the clitoris. In S. Hesse-Biber, C. Gilmartin, & R. Lydenberg (Eds.), *Feminist approaches to theory and methodology: An interdisciplinary reader* (pp. 301–329). Oxford: Oxford University Press.

Venn, C. (1999). Occidentalism and its discontents. In P. Cohen (Ed.), *New ethnicities, old racisms?* (pp. 37–62). London, New York: Zed Books.

Venuti, L. (1995). *The Translator's invisibility. A history of translation*. London, New York: Routledge.

Venuti, L. (Ed.). (2000). *The translation studies reader*. London, New York: Routledge.

Wekker, G. (2006). *The politics of passion: Women's sexual culture in the Afro-Surinamese diaspora*. New York: Columbia University Press.

Wekker, G., & Wekker, H. (1991). Coming in from the cold: Linguistic and socio-cultural aspects of the translation of black English vernacular literary texts into Surinamese Dutch. *Babel, 37*(4), 221–239.

Young, R. (1990). *White mythologies: Writing history and the west*. London, New York: Routledge.

5 Saving Masculinity in Cross-cultural Intersexualization

In 1979, New York endocrinologist Julianne Imperato-McGinley conducted research in two rural and remote villages in the Dominican Republic together with a research team. Imperato-McGinley shared Robert Stoller's working hypothesis that a biological force influences gender identity. In the same year, she also published a paper together with Stoller on one of his cases, which he attributed to the influence of his postulated biological force (Imperato-McGinley et al., 1979b). To test this hypothesis, Imperato-McGinley constructed a human research laboratory in the Dominican Republic to study the influence of testosterone (a hormone constructed as 'male') on male gender identity, using people she identified as 'male pseudo-hermaphrodites' to investigate her theories. Her account on intersexuality reveals an underlying heteronormative impetus that constructs femininity as pathology. In order to prove the influence of testosterone on the development of male gender identity (Stoller's coinage) in humans, she used male pseudo-hermaphrodites in a, what she calls, 'laissez-faire environment.' What emerges from her findings is yet another story of Western biomedicine imposing the figure of a pathological hermaphrodite onto a culture that had no such figuration prior to the interference of the researchers.

Several years later, anthropologist Gilbert Herdt, whose prior collaborations with the psychoanalyst Robert Stoller are discussed in Chapter 4 in this volume, continued his own research in Papua New Guinea, this time with the endocrinologist Julian Davidson. Herdt 'discovered' a third sex amongst the Sambia in Papua New Guinea, and suggested that Sambian culture has a corresponding third social category for gender. Herdt invited Davidson to collect blood and urine samples from people he identified as belonging to a third sex_gender, what was referred to as *kwolu-aatmwol*. Herdt's goal was to establish this third sex_ gender, not just on the basis of Stoller's psychoanalytical investigation and his postulated hermaphroditic gender identity mentioned in Chapter 2 in this volume but this time also on the collected bio-medical data. Herdt and Davidson originally set out to prove their hypothesis of biological 'male pseudo-hermaphroditism' as being a third sex as well as a third gender and thus hoped to find in Papua New Guinea the data that will support this. Imperato-McGinley's research team as well as Herdt and Davidson were interested in the diagnosis of 5-alpha-reductase deficiency, which falls under the nomenclature of male pseudo-hermaphroditism. Individuals are usually assigned to the male sex and will sometimes during puberty develop male secondary sex characteristics such as facial hair growth, deepening of the voice and no breast development.

Medical Anthropology and Neo-colonial Knowledge Production

Both research teams repeat the familiar formula of (neo-)colonialism that constructs so-called 'native' people and cultures as more natural, less restrictive, less cultured, and less civilized than their own. In both cases it is US-American culture that is considered as being non-naïve or more civilized and sophisticated. Both research teams show features that can be identified as neo-colonial modes of knowledge production. In the 1980s, a number of discussions occurred in anthropological debates on the usage of medical modes of interrogation and medical data. The historian David Arnold discusses in *Colonizing the Body* (1993) medical knowledge in terms of its contribution to colonialism. Arnold extends his analysis of medicine from a post-colonial and post-modern perspective, which according to him "cannot be regarded as merely a matter of scientific interest" and cannot "be abstracted from the broader character of the colonial order" (8). He states that "medicine was only one—albeit a particular critical—example of a whole colonizing process" (8).[1] This includes the exploitation of bio-medical data, which typically serves the knowledge production of the West at the expense of the people who are investigated. Bio-medical research is not undertaken to improve the health conditions of the sampled population. Its purpose is to increase the knowledge of the bio-medical experts of the Global North. When we learn about bio-medical lab research projects, we read the data and published findings, but we never hear the other side of the story. In this case, however, we have a rare opportunity. After Herdt's research in Papua New Guinea was completed, the Sambian people take the chance to talk back and demand compensation for the portrayal of their lives by Herdt.

Sampling-Naming-Creating: The *Guevedoche*

The 38 people in this study identified as having 5-alpha-reductase deficiency were from 23 families of four generations, some of them were related. Eighteen were raised unambiguously as girls of whom 17 changed into a male gender role during or after puberty. It is not noted how the sample was composed, or how individuals were identified as being male pseudo-hermaphrodites. Interestingly, Imperato-McGinley and her research team seem to have collected bio-medical data but it is not stated how and over which time period. In fact they primarily conducted interviews with individuals somehow known to be male pseudo-hermaphrodites and their families to obtain historical data. The individuals in the sample were diagnosed as having 'decreased dihydrotestosterone production due to 5-α-reductase deficiency.' 5-alpha-reductase type 2 deficiency is an autosomal recessive condition that prevents testosterone from converting dihydrotestosterone (DHT), which sometimes produces external genitalia for doctors not clearly identifiable as male or female. People are genetically male and have male gonads (testes) (see for example Wilson, 1999). Some anatomies are described as having 'hypospadias' (urethra opening on the underside of the penis) or a so-called 'mild clitoromegaly' or 'micropenis.' These diagnoses are of course only coherent with assigned sex at birth. There is no uterus or fallopian tubes but testes.

Imperato-McGinley and her research team published their investigation into the biological force in "Androgens and the Evolution of Male-Gender Identity among Male Pseudohermaphrodites with 5-alpha Reductase Deficiency" in the *New England Journal of Medicine* in May 1979(a).[2] Based upon psychosexual data gathered in two villages in the Dominican Republic, Imperato-McGinley et al. claimed that exposure of the brain to androgens "contribute'[s] substantially to the formation of male-gender identity" (1233). Imperato-McGinley et al. conclude that the surveyed and interviewed subjects demonstrate that "in the absence of sociocultural factors that could interrupt the natural sequence of events, the effect of testosterone predominates, over-riding the effect of rearing as girls" (1233). The researchers emphasize that their data "show that environmental or sociocultural factors are not solely responsible for the formation of gender identity, androgens make a strong and definite contribution" (1236). Therefore championing testosterone/androgens as a big influence on what they demarcate as male gender identity. The biological force that Stoller could only postulate has been given a name: androgen.

Imperato-McGinley states together with Robert Stoller in a collaborative publication on "Male Pseudohermaphroditism Secondary to 17 Beta-hydroxysteroid Dehydrogenase Deficiency: Gender Role Change with Puberty" in the *Journal of Clinical Endocrinology and Metabolism* in September 1979 that given the right conditions, androgen "can override the female sex of rearing" (Imperato-McGinley, Peterson, Stoller, & Goodwin, 1979b: 394). They find that, in the one case that Stoller has provided Imperato-McGinley with from his own sample at UCLA, psychosexual data was the same as in the Dominican Republic case. They argue one page earlier, somewhat more cautious that "some hormonal factors or biological force as previously postulated by Stoller, was strong enough in this subject to override the female sex of rearing" (393). In effect, the research team now supports Stoller's earlier hypothesis of a biological force (discussed at length in Chapter 2 in this volume) with endocrinological data.

Imperato-McGinley et al. conclude that "in the absence of sociocultural factors that could interrupt the natural sequence of events, the effect of testosterone predominates, over-riding the effect of rearing as girls" (1979a: 1233). In a nutshell, the argument is that testosterone/androgen is more powerful than social upbringing-as-girl for someone with testosterone deficiency. Interestingly, the research team does not clearly state, at what time the androgens/testosterone would become decisive—either *in utero*, during the neonatal period, or at puberty. Nevertheless, just as Stoller did before, Imperato-McGinley and her research team argue against John Money (et al., 1955) and the 'sexual neutrality at birth,' which they find in his theories (1235). By criticizing the research design of Money's study because the subjects were not matched for a similar hormonal milieu (they were matched chromosomally and gonadally), they intend to provide a solution for the issue of "nature (i.e., androgen) versus nurture (i.e., sex of rearing) in the determination of a male-gender-identity" (1235). In their endeavor to prove the influence of androgens on male gender identity, they refer to studies conducted on animals such as rats, guinea pigs, dogs, sheep, and rhesus monkeys in which the effects of androgens, given prenatally or/and post-natally can induce "both adult male sexual and non-sexual behavior" (1235). They state, however, that "in man, the question of the relative influence of hor-monal factors and environmental factors in the determination of gender identity remains unanswered" (1235). Nevertheless, they use the Dominican Republic case as a 'laissez-faire' environment in which the sex of rearing is supposedly "contrary to the testosterone-mediated biologic sex" in order to show that the biologic sex prevails if the normal testosterone-induces activation of puberty is permitted to occur" (1235).

Protofemininity Versus Masculinity as Achievement

Sex(ual) difference, neither in the psychological framework of Stoller nor in the endocrinological framework of Imperato-McGinley, is thought of in terms of symmetry; Stoller finds his theory of protofemininity and of the effort boys have to go through to become properly masculine backed up by Imperato-McGinley's theory of male hormones overpowering the female sex of rearing. Masculinity becomes *the* natural force on a psychological and biological level. It is con-structed as both, a hidden biological essence *and* a psychological achievement, which can literally 'overpower' and 'override' femininity. Biological in combi-nation with psychoanalytical argumentation is here used to reaffirm the status quo and the superiority of maleness and masculinity, and therefore consequently males and men. In Imperato-McGinley, Peterson, Gautier, & Sturla's research, the category of androgen/testosterone as a male/masculine hormone becomes intensely intertwined with psychological categories when she brings gender identity (Stoller) and gender role (Money) together with four so-called patterns of sexual behavior:

> sexual gender identity, sexual pattern (sex-related behavior, which for men includes direct aggressiveness, assertiveness, large motor activity and occu-pation), sexual object of choice (the sex of the person chosen as an eroti-cally interesting partner) and sexual mechanisms (the features of sexual expression over which an individual has little control, which for men include the ability to obtain and maintain an erection and to achieve orgasm).
>
> (1979a: 1234)

Androgen/testosterone here becomes the sole signifier of a specific behavior, which shows its effects at the onset of puberty. Here the connection between hormones—bodily substances—and social behavior such as aggression and profession is not just simplified but also binary sexed in the way that their effects are attributed to solely masculine behavior.

The fascination with the active production of male 'anything' (or 'substance') as an overwhelming force can also be observed in John Money's research, who wanted to champion rearing as the dominant force in the establishment of gender. Money's assertion that CAH[3] females show increased intelligence and a propensity toward lesbianism as a consequence of virilization is further proof of a patriarchal and sexist implication in intersexualization (Money, 1968: 40). Money also reports that "it is possible that the genetic factor responsible for CAH is linked to another genetic factor responsible for intellectual superiority" (40) thus implying that androgens are *the* cause for intelligence. Virilization as a biological process based on hormones and genetic dispositions is thought to be responsible for intellectual achievement—a rather commonsense assertion that reveals deeply sexist reasoning. Intersex activist and scholar Morgan Holmes has shown that the diagnosis of CAH and AIS as intersex conditions require a conception of hormones as sex(ualiz)ed_gendered.[4] Moreover, specific 'appropriate functions' have to be assigned in order to become regarded as diagnostic labels. Holmes states that:

> this is because the supposed masculinization seen in some females with CAH is attributed to an excess of "male" hormones in the female's body. Likewise, AIS is described as the "failure" of the chromosomally male (XY) body to respond to "male" hormones, and hence "fail" to develop the secondary sex features of a male.
>
> (2008: 48)[5]

Holmes stresses the asymmetry between the concept of maleness/masculinity and femaleness/femininity, which is found in the conceptualization of androgens/testosterone as male. Her conscious use of the vocabulary of 'failure' in regards to the body's ability to become male in some cases indicates the inherent patriarchal and sexist conceptualization of sex(ual) difference.

The individuals who did not conform to the presumed 'natural pull' towards masculinity are less favourably portrayed in the Dominican Republic research. One person is described as having "changed to a male-gender identity" but continues to "dress as a woman" (used to justify the category of "female-gender role"). Furthermore, it seems necessary for Imperato-McGinley et al. to mention that the person "has the affect and mannerisms of a man and engages in sexual activity with village women" (1979a: 1234). Another person has "maintained a female-gender identity and female-gender role" (1234). She was "married" (notice the inverted commas in original) to a man but was left by him after one year (1234). The depiction of this person is highly negatively connoted; the description reads as follows:

> She left the village, has been living alone and working as a domestic and has not been sexually involved with other men. She wears false breasts, yet

her build and mannerisms are masculine. She denies any attraction to women and desires surgical correction of the genitalia so that she can be a normal woman.

(1234)

The heterosexual matrix underlying these observations is also cultivated by stereotypical depictions of what is a real or normal woman. The imputation that she 'denies' to be attracted to women implicitly implies that she is or has to be according to her supposed biological masculinity. This reveals how heteronormative assumptions (with their focus on penetrative, vaginal intercourse), as well as assumptions about stereotypical gendered social behavior, shaped the study.

'Experiments of Nature'

Imperato-McGinley et al. draw the conclusion that in what they call a 'laissez-faire environment,' the sex of rearing when contrary to the testosterone-mediated biological sex, the:

biologic sex prevails if the normal testosterone-induced activation of puberty is permitted to occur [even though] the issue of nature (i.e. androgen) versus nurture (i.e., sex of rearing) in the determination of male-gender identity cannot be adequately resolved in these cases.

(1979a: 1235)

It remains unclear what a laissez-faire environment means; yet it implies that it is less culturally sanctioned or sophisticated. The core of the article, namely that the 'case' cannot fully be resolved, is contradicted by the following statement: This experiment of nature *[sic!]* emphasizes the importance of androgens, which act as inducers (in utero and neonatally) as activators (at puberty), in the evolution of a male-gender identity" (1236).

In 1981 another version of the article called "The impact of androgens on the evolution of male gender identity" was published in an edited collection (Kagan & Hafez, 1981). Both publications differ immensely in terms of attentiveness when it comes to a conclusion on the definite influence of androgens/testosterone on the development of a male gender identity. This time, Imperato-McGinley et al. state, that "it appears that the development of gender identity in man is continually evolving throughout childhood, becoming fixed with puberty" (1981: 106) and that the "time course for the development of a male gender identity with puberty appear to be unique to each individual" (106). Thereby relativizing their own findings from before. However, this does not seem to have any implication for the conclusions on the Dominican Republic case. Androgens are still connoted as the masculine essence, which can 'evolve' from a supposed non-evolved stage of femininity—the body leading the 'boy' into superior manhood that has always already been hidden in the form of hormones in the person's body.

Moreover, the term 'experiment of nature' implies here that this is nature but it is not normal for nature and therefore pathological. Hermaphroditism/ intersexuality has frequently been called an 'experiment of nature' but in the context of a neo-colonial setting, which we find in this case of medical research by US-Americans in the Global South, it achieves a new meaning in connection with the assumption that the culture of the other is a 'laissez-faire environment.' The supposed 'experiment of nature' is regarded as even more natural since it has allegedly not been influenced by culture: it has supposedly not been inter-rupted in its development, no cultural intervention is assumed to have taken place and thus it doubly serves in this logic as a test-case for the normal and natural. This is a common feature in bio-medical research in non-Western cul-tures conducted by Western researchers. Additionally, I argue that the failure of Western researchers to recognize that their pathologizing attitudes change the perception of people of themselves under surveillance. This is a significant feature of a (neo-)colonizing expansion. The semiotics of the West are super-imposed onto the 'other.' Imperato-McGinley et al. report that

> now that the villagers are familiar with the condition, the affected children and adults are sometimes objects of ridicule and are referred to as *guevedo-che*, *guevote* (penis (or eggs) at 12 years of age) or *machihembra* (first woman, then man).

> (1979a: 1235)

There is no indication that people were in earlier times distressed by the 'phe-nomenon' or perceived of it as such. Only after the researchers invaded the village did people think in terms of otherness about members of their com-munity. Through their investigations, Western researchers created a social cat-egory of sexual deviation that had not previously existed in the community. Those who do not follow the 'call of nature' that is the postulated biological force, are pathologized. Only after framing a so-called natural experiment as a 'phenomenon,' specific expectations and determinations can be put in place and those who do not obey their assigned appropriate designation are pathologized. Local understandings of sex_gender were overwritten and extinguished. The promotion of intersex* as an identity position, guarded and framed by the pro-cesses of intersexualization, has taken over the field of socio-cultural-symbolical orderings of this particular society. The same phenomenon is to be found in the study that has been done in Papua New Guinea only four years later.

'Male anything,' Endocrinology, and Aggressiveness

In feminist critical biology, especially Nelly Oudshoorn (1994) and Anne Fausto-Sterling (2000) offer the social and political explanation for how testo-sterone and androgen became 'male' and estrogen 'female.' According to them, designating certain hormones as male or female was fundamental to the project of establishing an essential and biological difference between male and female bodies. In the words of Morgan Holmes, Oudshoorn illuminates "the selective process through which scientists decide which truths they will discover" (2008: 41). In the case of hormonal research—as well as in the case of intersexualization in general—the truths to be discovered are the fundamental scientifically verifiable differences between the two sexes. Clearly sexed hormones serve not just in commonplace parlance to ground the dichotomous social organization of gender in a biological foundation. Anne Fausto-Sterling (2000) shows how scientists created the category of sex hormones and how hormones as such became markers of sexual difference and argues that if one looks closely and historically, it becomes clear that steroid hormones did not need to have been divided into sex and nonsex categories. She instead suggests that they could have been considered growths hormones that affect tissue and consequently reproductive organs and their function. In fact, so-called sex hormones, if one looks properly, affect several organs in the body and not just reproductive ones. The habit of attributing gendered meaning to body parts develops here a new meaning. Fausto-Sterling states that in the case of hormone research: "chemicals infuse the body from head to toe, with gender meaning" (147). Between the years 1889 and 1913, Fausto-Sterling identifies about 15 publications that influenced endocrinology in the sense that hormones became sexed and sexing. By the 1940s, science had identified, purified, and named hormones as sex hormones: "sex became chemical, and body chemistry became sexed" (158). Fausto-Sterling reports that in 1999 she found "300 articles mentioning estrogen and 693 discussing testosterone"—even though these nearly 1000 papers discussed a vast range of effects of estrogen and testosterone including Alzheimer's, nutrition, cancer, heart disease, the brain, the liver, muscles, and so on, they are still considered to be female and male hormones (179). Imperato-McGinley and her research team's findings testify to this history of hormones as infusing and being infused by gendered meaning—as well as to the masculinist asymmetry in assigning specific qualities to testosterone.

Feminist biologists have shown that there is a great deal of mythology especially about the effects of testosterone and other androgens and that studies show that increasing testosterone has no effects on mood, cognitive performance, libido, or aggression (Fausto-Sterling, 2000; Jordan-Young, 2010; Karkazis, Jordan-Young, Davis, & Camporesi, 2012). Karkazis et al. state that for the case of intersex* and bodily entities, fluids and markers:

> It is often assumed that people with intersex traits are somehow exceptional because of their complex biologies, but sex is *always* complex. There are many biological markers of sex but none is decisive: that is, none is actually present in *all* people labeled male or female.
>
> (6)

This, of course also accounts for sex hormones. Karkazis et al. here predominantly argue for the complexity of the biology of sex, which sheds light on the simplification that must go hand in hand with designating one category of hormones as male and the other as female. However, what also becomes clear is that the biology that is referred to in intersexualization is also very multifaceted and complex. This however, is not always acknowledged, not even, when it comes to the attempt of expanding the Western binary sex-gender-system from two to three. Gilbert Herdt's attempt to do so via research in Papua New Guinea fails; even though he might have started out to argue for multiplicity in sex and gender categories, the research he conducted together with the endocrinologist Julian Davidson repeats the faults of Imperato-McGinley et al.'s study.

The Sambia Turnim-Man

The paper "The Sambia 'Turnim-Man': Sociocultural and Clinical Aspects of Gender Formation in Male Pseudohermaphrodites with 5-Alpha-Reductase Deficiency in Papua New Guinea" presents clinical data from a joint field study conducted by anthropologist Gilbert Herdt and the endocrinologist Julian Davidson in 1983. Davidson was originally trained in classical neuroendocrinology and behavioral endocrinology in animals, and later extended his research to the relationship between hormones and sexual behaviors in humans. In 1963, he joined the faculty of Stanford University, where his research included the role of hormones in sexual arousal and behavior, sexual functioning in general, the influence of pharmacological agents on sexual functioning, and the social and cross-cultural aspects of genetic abnormalities. He, for example, was a founding editor of the journal *Hormones and Behavior* and co-edited *The Biological Basis of Sexual Behavior* (with Gordon Bermant, 1974). Herdt invited Davidson to join him in Papua New Guinea to study the so-called Sambian *Turnim-Man*. He hoped Davidson would provide endocrinological data to support the clinical ethnography that Herdt had already done with Stoller (see Chapter 4 in this volume) as well as to contest Imperato-McGinley's assumption that "sociocultural factors are not solely responsible for the formation of male-gender identity" and that "androgens make a strong and definite contribution" (quoted in Herdt & Davidson, 1988: 35). In fact, they state that their finding in the Sambia "provide a non-Western and naturalistic comparison" for Imperato-McGinley's studies.

Herdt and Davidson studied three generations of this "natural population" (1988: 42). They locate what they call "a small number of Sambia individuals" and provide psychosocial, behavioural, and hormonal data on individuals they diagnose with 5-α-reductase deficiency. They cite "intense and genetically close inbreeding" in this isolated population as one reason for the unusually high incidence of 5-alpha-reductase deficiency (42).[6] The rhetoric of an 'isolated population' feeds into the construction of a quasi-experimental setting that uses this population as a laboratory (Klöppel, 2008). In both the Dominican Republic and the Papua New Guinean studies, the researchers describe a 'natural' setting where intersex* appears in large numbers. They present both the setting and the population as 'natural.' The underlying assumption is the civilizing effects of 'culture' have not intervened in these places—implying that these populations did not have a culture. Just as in Imperato-McGinley's study, intersex* is understood as an 'experiment of nature' that is especially conclusive because it appears in supposedly native, untouched, isolated nature. The natural setting offers the researchers supposedly the perfect laboratory condition.

> The Sambia, an isolated tribe of interior New Guinea, provide an excellent example of a *naive [sic!]* population of 5-α-reductase-deficient males growing up and adapting to the traditional culture without Western medical intervention in their gender development and life-styles.
>
> (Herdt & Davidson, 1988: 52)

The traditional culture is presented as monolithic, unchanging, and naïve entity, presumably in contrast to the sophisticated and progressive West. Many of the familiar dichotomies that structure colonial logic (nature/culture, traditional/developed, isolated/interconnected) are present here. This construction of the other can be read as a process of neo-racialization or ethnicization, which functions through references to the natural and the traditional.

Herdt and Davidson claim that the Sambia "epitomize a three-sex category culture" and that "the gender differentiation of the pseudo-hermaphrodite cannot therefore fail to be ambiguous, for it is intermediate between the male and female categories" (Herdt & Davidson, 1988: 54) and they state that:

> Sambia recognize the existence of a third category, or third sex of person, the hermaphrodite *(kwoluaatmwol)*. It is this third sex that we must understand in its own cultural context, to avoid the ethnocentric assumption that all peoples the world over have but two sexual categories identified with the 'natural' sexual dimorphism of species, humans included.
>
> (37)

While this statement may be sensitive with regard to ethnocentrism, at the same time it suggests that this third category is somehow 'unnatural' by claiming sexual dimorphism is 'natural.' This makes cultures that recognize the "existence of a third" less 'natural' by default. Herdt and Davidson use psychosocial and behavioural data to support their bio-medical findings, but the actual blood and urine samples were taken from subjects from the Sambia population who show different psychosocial features and behavior. They admit that "detailed physical examinations have not yet been made by us" (34). Onto the sampled data they map an extrapolated diagnosis (44–50). However, the study was limited from the start since they only collected information from 14 people, eight of whom were dead. This of course reduces the blood and urine samples that could have been taken down to six. Of these remaining six, they were able to observe five, two of whom were children. The three blood samples they took indicated 5-α-reductase deficiency. They describe the two children as ambiguous in appearance, albeit more male-like in behavior, also as more "passive and quiet" (45), and one of them as behaving in a "boyish way" (44). The four adult subjects—noted as being larger than normal females—lived as women. Two are described as "never having inseminated boys" (49, part of the semen practices, see below) but one of them is an enthusiastic fellator and "persisted in homosexual activity" (refers to a ritualistic semen practice, see below). A third adult "likes to go out with women" (48) but to Herdt and Davidson's knowledge does not engage in hete-rosexual intercourse. A fourth adult "uses prostitutes" (49). Interestingly, all the psychosocial and behavioral descriptions of the five subjects are (metonymi-cally) reduced nearly completely to assessments of their sexual activity.

Although Herdt and Davidson criticize Imperato-McGinley's research for being essentializing, their own theory of the natural tendency of the hermaphrodite towards maleness is equally so. In "Mistaken Gender: 5-Alpha Reductase Her-maphroditism and Biological Reductionism in Sexual Identity Reconsidered" (Herdt, 1990), he writes that "cultures such as our own, which overlay sexual dimorphism in nature upon gender identity development in humans, tend to be essentialist and morally restrictive regarding conceptions of personhood and sexual conduct" (435). Herdt implies a clear distinction between nature (sexual dimor-phism) and culture (gender identity) for Western societies. He describes the Sambian culture as "less restrictive in socialization and more accepting of sexual variations, making of androgyny a significant motif in cultural representations, even in the sacred" (435). In light of the 1980s publication date, the commentary on androgyny is noteworthy. In the USA, androgyny abounded in cultural repre-sentations, even if only in pop-culture. Herdt disregards this to create a dicho-tomous relationship between the two cultures. By producing hermetic and binary portrayals of differences between the USA and the Sambian culture, Herdt essen-tializes these differences and establishes them in relation to the framework of androgyny. Androgyny, derived from Greek *Andros* = male and *gyné* = female prevents the possibility of exiting the binary symbolic organization of Western sex_gender and consequently makes representation of the 'less restrictive' and 'more accepting' culture difficult if not impossible. The 'mistaken gender' in the article title refers both to the *guevedoche* (in the Dominican Republic) and the *kwolu-aatmwol* (in Papua New Guinea). The 'mistake' for Herdt is that the indi-viduals were assigned as female and not third sexes_genders (1990: 442).[7] Because the "three-category system provides for greater fluidity," he says the hermaphro-dite can "switch from mistaken female-defined to male-defined hermaphrodite." He proceeds as follows: "Only a profound inner sense that one is inexorably female would inhibit such persons from making the structural sex transformation from exposed 'female' to hermaphroditic male" (442). Herdt understands this pro-found inner sense, by referring to Stoller (1975) as "an identity state similar to that of the primary transsexual" (442). He thereby naturalizes the third gender as intrinsically male, since he installs a single-caused—i.e., biological reason for the shift to a male identity position. The question remains why Herdt so insistently criticizes Imperato-McGinley's research as essentializing if he does exactly the same via the supposed natural tendency of the hermaphrodite towards maleness. However, in this paper, he still refers to 'mistaken gender.' In an article published four years later the only significant change is that Herdt replaces 'mistaken gender' with 'mistaken sex.'[8] A further difference of these two versions of the paper consists of Herdt's earlier use of *gender identity* (1990: 443), which is replaced by the term *gendered socialization* (1994: 444). The rest remains the same, as it states that both terms refer to the hermaphrodite as being "not unambiguously male or female" thereby in contemporary discourse implying sex. Merely, in the version from 1994 a sentence of interest follows: "the cross-cultural variations reviewed in this essay attest to the importance of gendered signs of identity as cultural and historical achievement, with implications for the emergence in certain times and places, of a third sex" (444). In both papers, Herdt interrogates sex assignment and socialization in what he calles three-sex-cultural systems. Herdt states that he refutes the "unicausal biological model" to suggest that psychocultural factors in these cultures are prevalent in the development of gender identity. However, his uses of the categories of sex and gender are highly complicated in terms of an essentialization of the former at the expense of the latter. Initially, Herdt agreed that the cultural construction of a third sex, the *kwolu-aatmwol* in the Sambia, is "inexorable" (Herdt, 1981: 434). However, Herdt states that continuous field study has made him apprehend "that while Sambia recognize three sexes and at birth sex-assign them as such, their world view systematically codes only two genders, masculine and feminine in cultural discourse" (Herdt, 1990: 434). This is rather puzzling, since the following conclusions contradict this statement. Herdt refers to Stoller's *Sex and Gender* (1968), where he argued for the "primacy of sex assign-ment in the determination of gender identity" and consequently sees "anomalies here with regard to ambiguous sex assignment: that is, the possibility of hermaph-roditic gender identity" (Herdt 1990: 434). Herdt finds this "hermaphroditic gender identity" in the Sambia. He acknowledges that "cultures such as our own, which overlay sexual dimorphism in nature upon gender identity development in humans, tend to be essentialist and morally restrictive" (435). Yet still he constructs an overlay of sexual tri-morphism upon gender identity. In his effort to refute Imperato-McGinley, he writes:

> Contrary to the bio-medical explanation, then, my hunch is that the Domini-can guevedoche does not experience postpubertal developmental change as being from "female to male". Instead, the transformation may be from "female"—possibly ambiguously reared—to male-identified hermaphrodite, who is, in certain social scenes, categorized with adult males.
>
> (Herdt, 1990: 438)

Herdt refers to a historical/symbolic category in Sambia that suggests parents and midwives know that some children are born with "anatomical ambiguity" and will experience "dramatic masculinization" at puberty (439). He concludes that "hermaphroditic infants are sex-assigned as *kwolu-aatmwol*, not as male" (439). This third gender position is reserved for "male pseudohermaphrodites" who live in a male gender role.

> In the four historical cases of sex role change, the female-defined as kwolu-aatmwol did not convert to a different role until after their exposure and

failure as female. One of them still lives as female. In other words: social catastrophe forced them to change, or else face an unbearable and ambiguous future as not longer clearly female but not yet male-associated pseudohermaphrodites.

(1990: 441)

Herdt later states that his hypothesis of the third sex category is proven by the assignment of hermaphroditic infants as *kwolu-aatmwol* and not as male; as such, infants who are assigned as female are 'mistakenly' reared as 'normal' females.

Yet because the phenomenon has existed for generations and the midwives and mothers go to some lengths to examine the infant's body for signs of the *kwolu-aatmwol*, it is unlikely that a mistake in sex assignment will occur, and only a few instances of such a mistaken assignment, (…) are historically known.

(1994: 436)

This argument can be read as supporting the perception that there is an initial 'true' male unmistaken sex assignment because the assignment as female would be mistaken. The third gender is conceptualized as a tendency to maleness/ masculinity. Therefore, the descriptive terminology is marked with 'effeminacy.' Morphological features as well as sexual behavior are measured against heterosexual masculine/male standards, whereas female/feminine aspects are disregarded as irritation. Herdt consolidates Stoller's 'male hermaphroditic gender identity' and secures masculinity/maleness, using terminologies of effeminacy. His third sex/third gender thereby attains a 'natural' tendency towards the 'first' sex. It is first *andros* and then some *gyne*. This consolidates not just Herdt's but also Stoller's bias against femininity. If read in line with Stoller's theories on the biological force as well as his psychoanalytical theories on protofemininity— masculinity is an achievement, not just in so-called male pseudo-hermaphrodites but in humans in general whereas femininity is the generic and basic form of either biological of psychological becoming.

Origin Myth Versus *Turnim Man*

Herdt and Davidson compare the three-sex system to the two-sex-system. Whereby even though the researchers attempt to allow for diversity, they homogenize the parameters according to which they measure their data.

The causes of pseudohermaphrodism are a mystery to the Sambia. They do not understand why some people are born with this condition, nor do they hold explicit beliefs about its causation. In a secret origin myth of parthenogenesis, however, the men view hermaphrodism as primeval foundation of maleness or femaleness.

(1988: 38)

In the now-not-so-secret-anymore Sambia origin myth, (conveyed by Herdt), the ancestors *Numbugimupi* and *Chenchi* were "two beings of hermaphroditic nature, a blend of male and female anatomy, whose sexual interactions impregnated one and masculinized the other, thus founding society, but also initiating the secrecy on which male power and warfare were based" (Herdt, 2003: 109). Herdt interprets this as "homoerotic oral sex and then coitus, leading to birth and the founding of the society" (109). For Herdt, this explains the creation of the genders "out of the original hermaphroditic state of humankind" (109). Genders are created out of the hermaphroditic state through a twenty-year process of rituals during the life cycle of male members of this society. Hermaphroditism is embedded in the origin myth. To assume that they do not understand its causation is to superimpose the Western explanation system of bio-medicine. Does bio-medicine have an understanding of why hermaphroditism exists? Western bio-medicine, in fact, only knows that the dichotomous system of two sexes is constructed and that surgery is used to correct, surgically modify, and erase such bodies, which has been constructed as abnormal.

In recent years, the Sambia adopted from the Pidgin trade language the term *turnim man* (*kwolu-aatmwol* is the indigenous term). Herdt & Davidson find that *turnim man* is:

> apt because it emphasizes (i) the "process-of-becoming" quality, and (ii) the feeling that these ambiguous anatomical beings are driven biologically to be more male-like persons. The Sambia do not understand this biological drive: They think that it is "natural" for these people to transform but they do not believe that they will become normal, unambiguous masculine males. This is why they do not call them "men" but rather turnim men: The Pidgin term connotes better than their own indigenous one this transformational "biological drive" attribute. That this is so can be seen from the fact that the turnim-man concept is widely used by everyone—men, women, and children—most of whom speak no Pidgin but have simply taken over the concept into their vocabulary slang.
>
> (1988: 38)

Herdt looks over the creation myth and prefers the *turnim man* model because it aligns with his theory of the biological drive or impulse.[9] In Herdt's coordinate system the *true sex* of a hermaphrodite has to be male pseudo-hermaphroditic because Davidson has diagnosed 5-alpha-reductase deficiency as the bio-medical intersex* condition. Thus Herdt prefers the *turnim man* because it helps his claim that male hermaphroditism results in male gender identity. The *turnim man* can only be healthy and normal if he adopts male gender identity. Thus a male biological pseudo-hermaphrodite winds up with a male gender identity. The 'process-of-becoming,' which is inherent in the term, is exclusively readable in relation to the biological imperative to maleness/masculinity. I suggest that this should be put differently: the mature and healthy Sambian *kwolu-aatmwol* adjusts through becoming the *turnim man* to Western standards of the heterorelational sex-gender-sexuality-system in which every individual has only one sex and the corresponding gender—even if it is framed hermaphroditic. The Sambian *kwolu-aatmwol* has been colonized by Herdt's various translation efforts and by the Pidgin. In *Colonial Desires* (1995) Robert Young identifies Pidgin and creolized languages as powerful models for the analysis of colonization since "they preserve the real historical forms of cultural contact" (5). This is not about grammar or the abstract debate in colonial studies about how power operates. He continues, "the structure of pidgin—crudely, the vocabulary of one language superimposed on the grammar of another—suggests a different model from that of a straightforward power relation of dominance of colonizer over colonized" (5). The team in Papua New Guinea had the same effect as the one in the Dominican Republic. Both imposed truth claims and caused deleterious material effects. One of the children said he "felt 'shame' being interviewed" (Herdt & Davidson, 1988: 45). "This is understandable," Herdt and Davidson admit, "since his peers knew he had been singled out as a hermaphrodite in the research" (45). It seems rather astonishing that they do not question their approach after they found out that this is one of the consequences of their investigation. The identity position they might have found in the *kwolu-aatmwol* is one that has traditionally led for example to shamanic training and duties. After their interference the *kwolu-aatmwol* is connected with shame. The slogan of the 'end of shame and secrecy' was one of the first ones in the US-American intersex social movement.[10]

The Problem of Translating Identity

The difficulty of translating *kwolu-aatmwol* is particularly challenging if we take into account that identity categories in Melanesia are not Western. Anthropologist Deborah Elliston faults Herdt's flawed assumptions about Melanesian identities, which are, according to her:

> far more flexibly constructed than in Western societies. The Western 'identity' construct requires radical qualification for Melanesian societies, as durable for a time period much shorter than a lifetime, and as meaningful in relation to the exchanges of substance, not in relation to an essentialized and internally consistent, individuated core persona.
>
> (1995: 853)

In Melanesia, the distinctly non-Western boundaries between men and women require an extensive amount of work (rituals and rules) to maintain. It would be a "mistake to construe these conceptual boundaries in Melanesia as similar to Western boundaries between gender categories" (853). This concept of masculinity and femininity differs from Western notions because these notions require to be produced by social interactions such as rituals; as such, they are not perceived to be caused by a biologically grounded morphology (or in our case a biological force), which determines a person's identity for life. Masculinity and femininity seem to be processes rather than fixed states of being; they need to be confirmed time and time again and they change in their meaning and value according to age and status. This derives partly from the various gendering processes that anthropologist Marilyn Strathern (1988) has described in the Melanesian culture, namely that "the gender of people's sexual organs depends on what they do with them" (122). As such, the emphasis in gender construction lies here on the type of relation or interaction in which the genitals (or sexual organs more generally) are employed. Moreover, Strathern notes that "the unitary identity sets the stage for the revelations that it covers or contains within itself other identities" (122). In fact, she states that it could not be further from the Melanesian case to assume that there is "a single-sex social identity that finally matches an intrinsic physiological one" (123). Moreover, Herdt's problematic term of ritualized homosexuality has also been dismantled as the imposition of Western models of sexuality to explain Sambian 'semen practices.' Elliston says Herdt wants "to minimize the assumptions that can be imputed to the semen-focused techniques and ordeals through which boys are 'made into men' in some Melanesian societies" (1995: 850). She suggests the these rituals aren't best understood as sexual and discusses the biological explanations of semen practices stating that these explanations "may be most interesting for what they reveal about Western models of biology and homosexuality" (851). Identity and personhood as Western concepts are also discussed by Elliston, as she juxtaposes these with Melanesian concepts of personhood, which are "arguably far more flexibly constructed than in Western societies" (853).[11] Strathern also notes that "concern with identity as an attribute of the individual person is a Western phenomenon" (1988: 123).

The above-mentioned revealing statements by Elliston apply also to the third 'hermaphroditic identity' Herdt, Stoller, and Davidson have claimed. 'Identity' per se is a Western concept, which is supposed to be intrinsic to a person and stable through life: not acquired but innate. The term sex is similarly problematic since it implies an entire discursive apparatus that has brought sex into being as the 'biological truth' of a person's identity (see also Foucault, 1978). Moreover, it is an attribution of "sexual significance to what the participants may under-stand in many other ways," not just in sexual terms (Hoad, 2000: 149). Identity as a category of self is in the Western sense intrinsically tied to its sex-gender-sexuality-system; however, it may not be the case in other cultures (see also the discussion in Chapter 4 in this volume).

Social anthropologist Henrietta Moore, in a psychoanalytical reading of Herdt's material, which I find rather limited in its rewriting of the Oedipus complex, does however provide some interesting statements on the Sambian construction of the self in relation to non-Western kinship systems. Moore explains convincingly that the symbolic system in the Sambian world involves "an exchange of objects that is constitutive of the psychic and social processes that make up the body image, self-other relations and gendered identities" (2007: 155) and concludes that:

sexual identities and the ideas of masculinity and femininity which sustain Sambia understandings of sexual difference cannot be easily reduced to the forms of 'individuality,' with their concomitant ideas about autonomy and separation, which are often dominant in the West.

(156)

Therefore, to adapt the concept of the individual to the Sambian notion of identity, or perhaps rather self or subjectivity, seems already to be a mistake. The Western notion of sex being the truth of the person is implicit in the cross-cultural ethnographic research that is conducted by Herdt, Stoller, and Davidson. A notion which Moore tackles here as probably inappropriate for the societal organization of the Sambia; at least, she questions whether sex might be related to identity in the same way as in Western narratives. Therefore, by replacing the origin myths by one single psychological and bio-medical explanatory model, a colonization of another symbolic and lived system of social relations takes place. As Herdt remarked in *The Guardians of the Flutes* (1981), the Sambia "make no distinction between sex and gender—the sexed body and the culturally and socially constituted identity" (168). Moore expands on this in her psycho-analytical reading of the material on the Sambia and elaborates on the ways of becoming a sexed being in the Sambian culture:

> A Sambia man lets blood from his nose in the occasion of each of his wife's periods, thereby linking the maintenance of his masculinity to a cyclical process that rids him of the contaminating effects of femininity. If the male and the female were sufficient, this would not need to be so. Sexual difference is the retroactive consequence of the effects of masculinity and femininity, and of their careful management within symbolic systems and social relations. The irony here is that sexual difference is maintained through its impossibility through the failure of representation to capture in any lasting and fixed sense the distinction between masculinity and femininity. Nose-bleeding reveals that masculine subjectivity is founded on that impossibility which becomes both the 'how' and the 'why' of sexual difference itself.

(164)

The point Moore wants to get at here is that "there is no problem about males and females, but there is a problem about masculinity and femininity" (160) and that "the irony is that it is gender that creates the problem of sexual difference: it accounts for the necessity of the latter's repetition; gender is why the male and the female are never enough" (160). I disagree with Moore's use of the terms of sex(ual) difference and gender because she affirms a fundamental distinction between the two and in her vocabulary there is no exit from the sex_gender distinction. Yet, I regard her analysis of Herdt's material on Sambian culture in its focus on the effects of masculinity and femininity as enlightening. The consequence that can be drawn from this is that sex(ual) difference never exceeds gendering processes and that different experiences of embodiment cannot be reduced to a Western explanatory framework.

In Herdt's account, however, the Sambian sense of bodily experience, which is grounded on the origin myth and a specific form of embodiment, is disregarded and modified according to Western theories of sex-gender-sexuality-systems. In order to justify his hypothesis of the third hermaphroditic gender identity, Herdt exclusively draws on one case—the one of Sakulambei (Chapter 4 in this volume), which seems very different to those of other *kwolu-aatmwol* mentioned above in the paper with Davidson. The *kwolu-aatmwol* can neither be translated into Western discourse nor is the 'identity' or position of the different *kwolu-aatmwols* in Sambian culture the same. The *kwolu-aatmwol* cannot be unified as a category, neither from a Western perspective nor in the Sambian symbolic organization.

The philosopher Emmanuel Levinas coined the term 'egology' for the subject-centered philosophy of the West (Levinas, 1998). He explains this term as referring to the legitimization of the exclusion of the other, whose otherness is performatively produced as part of its own speech. Sociologist Couze Venn, drawing on this concept argued that:

> the connection with occidentalism is that in the space opened up by colonial conditions, egology found its proof and its measure. At the same time, its discourse, from Descartes, becomes progressively secularized, vested in a specific notion of rationality; it is naturalized in the discourse of Darwinism.
>
> (1999: 51)

The colonization of the self of the other in terms of Western egology is based on the conceptualization of the self as individual and not as connected to other selves. Western discourse is unable to understand embodiment as independent from the concept of identity, which, in the case of Herdt and Davidson, becomes a bio-medical identity. Yet this is not to say that one singular notion of the individual in the West exists either. Anthropologist Melford Spiro in his interrogation of the Western conception of the self as "peculiar" states that "bipolar types of self—a Western and a non-Western—are wildly overdrawn" (Spiro, 1993: 116) and that "surely, some non-Western selves, at least, are as different from one another as each, in turn, is different from any Western self" (117). Multiple meanings, lived experiences, embodiments, and subject(ivitie)s are, in accounts on sex(ualiz)ed_ gendered identity and the like, simplified under this one notion of the individual. Conceptualizations of self vary according to context and modes of expression. However, the hegemonic concept in discourse implies a unified version of this Western self; the coexistence of different dimensions of subject(ivitie)s are ignored.

Anthropology and Neo-colonialism: Material Effects

Gayatri Spivak in her famous article *Can the Subaltern speak* (1988) accuses Foucault and Deleuze of neglecting *epistemic violence* because they are, even though conscious about the connection between discourse and power, projecting a white European epistemology onto the rest of the world, thereby producing gross universalizations. For Spivak, postcolonial studies are also exerting epistemic violence since they ironically reinstall, co-opt, and reinvent neo-colonial imperatives of economic exploitation, political domination, and cultural and symbolic as well as semantic erasure. Post-colonial theories, so Spivak's contention, are complicit in creating a neo-colonial epistemic discourse which is inevitably violent since it again silences the other by working with and referring to institutionalized discourses. In turn, these discourses classify and represent the other in the same manner as the discourses and material effects of the colonial asymmetry they seek to dismantle. Any attempt from the 'outside,' that is—the West, to ameliorate the colonial condition of the other by 'granting' them collective speech invariably will encounter the problems of first a logocentric assumption of cultural similarity among heterogeneous people, which happens via the construction of the other as homogenous and second a supposed dependency upon Westerners to 'speak for' the subaltern rather than hearing them speak for themselves. The assumption of a subaltern collectivity—the homogenous other—is again reproducing the ethnocentric extension of Western logos—the totalizing, essentialist 'mythology' Derrida (1982) has described, which fails to account for the heterogeneity of the colonized 'other.' The Western logocentric myths cannot account for the symbolic, cultural, historical systems nor their body politics or their specific conceptualization of subject(ivitie)s or even resistance to all of the aforementioned. Even though Herdt's work cannot claim to be post-colonial; as I argued, it is rather neo-colonial, I suggest, that he is exerting this kind of epistemic violence which Spivak talks about. The following instance of a communication reported in an important anthropological journal testifies to this epistemic violence, yet also to the material violence which Herdt's work exercised over his 'interpreters.'

In the journal *Anthropology Today* from August 1998 (14(4), 30) a report appears in the miscellaneous section under the rubric 'Media,' stating that a letter was sent to one of the two major international newspapers of Papua New Guinea, namely the *Postcourier*. According to the report, this letter was signed by an Eastern Highlands people who protest against their government "allowing a well known American anthropologist to study their culture in the 70s and 80s" (30). The letter states that the author's book is against their customs and that the author did not leave copies of his book at the Institute of Papua New Guinea. The report furthermore states that court action has been taken against this author to claim compensation "for the damage his books *are alleged* to have caused their customs and traditional beliefs" (30 [my emphasis]). The author of the report adds that

> the implications of the raising of this issue in the context of "compensating culture" of Papua New Guinea, and the growing awareness of indigenous rights and the power associated with their recognition throughout the Pacific (and elsewhere) is hard to overstate. *A large can of worms may be opening...*
>
> 30 [my emphasis]

In the same journal in February 2000 (16(1), 26) in the rubric News Follow-Up we find the note that a "similar letter *purporting* to be from eleven clan representatives has now been received by the American Anthropologist (AN)" (26). This letter was then published together with a response by Herdt, whose identity has thereby been revealed. Herdt in the process has "prevailed upon AN to replace the name of the group used in the letter with 'Sambia' " (the pseudonym he used) in order to "maintain the anonymity of those whose homoerotic rituals" he has described "despite the compensation claimants' complaint that the term [ritualized homosexuality] is derogatory" (26). Herdt, in his response letter, points out that he has been warned of "unsolicited letters from friends in Sambia" by "unscrupulous efforts of several unemployed Sambia (..) to extract compensation payments" (26). Herdt furthermore states that all anthropologists are vulnerable to "accusations of this kind, which may seriously imperil the rights of both anthropologists and the people they write about" (26). Both reports (unfortunately I could not get hold of the letters) are written in an insinuating tone. I have emphasized all the words, which silence the people who speak from a subaltern position. The terms *alleged* and *purporting* are used to devalue first the claims of compensation, second the cause, and third the speakers. The fact that Herdt has not left copies of his books in Papua New Guinea speaks for itself and supports my analysis of the profit which Westerners have gained via the knowledge they have produced about 'the others' while not letting them participate, neither in the production, nor the consumption of knowledge about their own lives. The mentioning of the "unscrupulous efforts of several unemployed Sambia" furthermore serves to question the integrity of the speakers by insinuating that they are first 'losers' and second only after the money. That Herdt asked the journal to replace the name of the people whom he researched and who now claim an authentic position to speak with the pseudonym he gave them is a further silencing; this, again, implements the authority of the anthropologist. The fact that what Herdt has called 'homoerotic rituals' is perceived by the claimants to be a derogatory term is not addressed, which highlights the gap between the anthropologist's categorization and interpretation and the lived experience of the people studied. The statement in the report that "*a large can of worms may be opening. . .*" (26 [my emphasis]) is an inadequate phrasing in the first place, because it evokes a picture of the Papua New Guineans emerging from their confined and preserved stage into the world, questioning the authority of the anthropologists, and furthermore unrightfully claiming rights, power, and compensation.

Anthropology as a discipline helped significantly in the production of knowledge on colonized people in order to administer them. Ethnographic and anthropological knowledge provided systematic information about the colonies for the colonizers to be able to effectively rule them. Cultural anthropologist Jack Stauder interprets the development of anthropology "in terms of the material situation in which it was practiced" (1993: 417). The main shift in the colonial geo-political complex is that the United States has replaced Britain as the leading imperialist power in the world. The form of colonizing agendas has changed but not their content. Stauder emphasizes that contemporary anthropological research is as much as any kind of scientific research "systematically shaped and utilized by the dominant interests in our society" (425). Another agenda, besides the one of the colonization of bodies, is the production of knowledge through the appropriation of the bodies of others to enhance Western scientific data. This abstract form of profit achieves its explicit shape when the people under investigation are sampled, tested, categorized, disciplined, normalized, and subsequently pathologized. In the context of colonialism, medicine has a clearly specific function and always had a "place within a more expansive ideological order and a wider empirical domain" (Arnold, 1993: 8). The dominant interest in the cases described above seems to be the dissemination of knowledge about the other for the West and not for the other. The material colonial situation is here relived and reproduced by the silencing of the other and their claim for compensation.

In the 1980s, during the time of the studies analyzed in this chapter, a number of theoretical discussions were brought up explicitly in anthropological debates on the usage of medical modes of interrogation and medical data. In 1986, John de Cecco describes the swing back to the biological theories of the early twentieth century in the anthropology of (homo)sexuality (1986: ix). He sees a biological etiology and a biological reductionism at play in anthropological studies of homosexuality and asks for a "psychological detoxification" and an increase in studies that show the "many cultural faces of homosexuality" (ix). This quest can be easily adapted to the processes of cross-cultural intersexualization. Biomedical research in a neo-colonial context is laden with power and cannot be conceptualized without the political context in mind (e.g., Fanon, 1965). David Arnold states that "the accumulation of medical knowledge about the body contributed to the political evolution and ideological articulation of the colonial system" (1993: 8). Arnold extends his analysis of medicine from a post-colonial and post-modern perspective, which according to him "cannot be regarded as merely a matter of scientific interest" (8). Medicine, according to Arnold, "cannot meaningfully be abstracted from the broader character of the colonial order" (8). In cross-cultural intersexualization this broader character of what I call the neo-colonial order is obvious; pathologizing moves which take place here are violent in an epistemic way. Clinical anthropology, or in this case, clinical ethnography (see Chapter 4 in this volume) conducted by Westerners led to a scientific colonization or anthropologization of specific areas of the world and its inhabitants (Foucault, 1994; Knauft, 1999).

Today it is acknowledged that anthropology and ethnology/ethnography descend from a tradition of travel reports and the like. Moreover, anthropology arose largely out of the need for colonial administration (Forster, 1973); therefore, a certain synergy between economical interests and anthropology cannot be denied (Grosse, 2000). The origins of anthropology and its methodologies are thus colonial, having emerged in an era when the world was mapped out geographically and politically. This era displayed a particular arrangement of power relations and hierarchies embedded in the expansion of European empires. Colonialism should be regarded as not just a matter of military invasion and economic exploitation; it should also be seen as a practice of imagination through which dominated populations are represented in ways that produce ethnicization/racialization and in this case cross-cultural intersexualization.

Conclusion

The studies by Herdt, Stoller, Davidson, and Imperato-McGinley analyzed in this chapter represent the Western expansion of essentializing, normalizing, and pathologizing structures in knowledge production. They reveal that even the scholars intent to destabilize biological essentialism as in the case of Herdt—and his third sex or gender—the (neo)-colonial framework counteracts. What was supposed to help multiply notions of sex and gender, helped only to re-mediate and re-construct the hierarchical 'natural sex(ual) difference' between man and woman. This, as I hope to have shown was only possible through the newly discovered ontology of hermaphroditism/intersexuality in other cultures. The syndrome of 5-alpha reductase deficiency, which describes 'intersex conditions' was applied to interpret social behavior, such as change in appearance, interest in certain working duties, and so forth. Biological forces to which the subject has supposedly been exposed before birth are brought into motion to explain gender identity as a psychological category (see Chapter 2 in this volume). The implicit masculinist bias that can be detected in Herdt's, Davidson's, Stoller's, and Imperato-McGinley's work only manifests the pathologization of femininity at a universal and generalized level. The male pseudo-hermaphrodite is categorized on the basis of his biologically induced masculinity and is subsequently normalized if the third gender conforms to this masculinity. The expectation that endocrinological data will prove that testosterone produces heterosexual, masculine behavior underpins both studies. Their perception of male pseudo-hermaphroditism requires the way they interpret their interviewees' behavior; as such, they pathologize those who do not live according to standards of masculinity. Consequently, the postulated third sex and/or gender, therefore, is not the third as a neutral position in the two sex_gender system but always leans towards masculinity.

The epistemic power the studies reviewed here exert is embedded in a colonial setting and an extension of Western bio-power and symbolic violence. Herdt and Davidson assume that the self is constructed similarly in the Sambian culture as it is in the West; the *kwolu-aatmwol* cannot be heard by them and neither can they integrate the origin myth of the Sambia in their interpretation since they do not regard it as scientific. The invasion and colonization of the semiotic and symbolic system of the Sambia via the new Pidgin term *turnim man* replacing the *kwolu-aatmwol* (as well as the newly coined term of *guevedoche*) mirrors the appropriation of the multiplicity of meanings and lived realities. The question of who has the authority to speak and what counts as valuable and legitimate knowledge is in this framework just too obvious. The semiotics as well as the symbolic system of the (neo-)colonizers is superimposed onto the colonized and is supposed to exhaustively explain their cultural and symbolic systems. This means that a male sex, a female sex, and a hermaphroditic sex become the preconditions for the existence of a masculine identity, a feminine identity, and a hermaphroditic identity. Sex—as a now threefold biological category—becomes sexual identity and replaces gender. The third is bound to the supposedly trimorphic 'natural' and universal make-up of sex, gender, and sexuality. By disciplining anything that might interrupt the clear-cut categories of interpretation, the original order is reinstalled only with a new taste to it: everything can be included in the Western system of explanation; there is nothing that cannot be grasped with the system of bio-medical knowledge production.

In this chapter I looked at Herdt's claim of the necessity to invent a third category, not just for humans but also for animals. He states that "as a cultural ideal this category may be perceived and projected into the order of nature" (Herdt, 1990: 442). The question that arises here is the following: Is it necessary to project a new category into the orders of nature if one wants to emphasize that orders of nature are subject to cultural interpretation anyway? Aren't exactly these orders of nature the problem, which prevent a reevaluation of the cultural imperative of the two sexes_genders? Moreover, the third, as an empty signifier, has developed a life of its own traveling the world as I will explore in the Excursus and Chapter 6 in this volume.

Notes

1. Arnold speaks here of the treatment of epidemic diseases such as cholera, smallpox, and the plague in colonial India. His account focuses on the practices that have been applied during these epidemics. He also stresses the point of the unequal yet mutual relationship between local medicines and Western medicine and analyzes the colonizing habits of Western medicine in its historical, political, and cultural dimensions. He stresses the institutional and technological dominance by the British colonizers over the colonized because in India, Western medicine was applied and it also served to manage the colonies. However, the medical knowledge was sometimes refashioned in response to local needs.
2. Their study came from a clinical investigator-award from the National Institute of Arthritis, Metabolism, and Digestive Disease, as well as grants from the Clinical Research Centre, the National Institutes of Health, the National Foundation–March of Dimes, the Shorr Fellowship Fund, and Gulf and Western This article is quoted in psychology and sociology textbooks such as R. D. Gross (1987). *Psychology: The Science of Mind and Behaviour,* London: Hodder and Stoughton.
3. Congenital adrenal hyperplasia (CAH) refers to excessive or deficient production of sex steroids. It can alter the development of so-called primary or secondary sex characteristics. The definition as an intersex condition varies according to which symptoms are included in the definition of intersex.
4. Androgen insensitivity syndrome (AIS) is commonly described as a "genetically inherited change in the cell surface receptor for testosterone." CAH is described as an "inherited malfunction of one ore of more of six enzymes involved in making steroid hormones" (Fausto-Sterling, 2000: 52).
5. Morgan Holmes furthermore adds to this that "many persons with an XXY karyotype (Klinefelter's syndrome) or with CAH or androgen insensitivity syndrome (AIS) could not have been diagnosed as intersexed prior to the late 1950s" (Holmes, 2008: 48).

6. Inbreeding as a cause for intersexualized birth is a popular trope for medical explanations.

7. The term *third gender* in relation to cross-cultural research appeared first in 1975 in a feminist-motivated anthropological monograph (Martin and Voorhies 1975). In the chapter "Supernumerary Sexes" in *Female of the Species*, Kay Martin and Barbara Voorhies employed the category of *third gender* to draw attention to the anthropological evidence that the two-sex/gender framework of the West could not adequately explain the organization of other cultures. In 1975, the *gender-concept* and the implicit distinction between *sex* and *gender* had already been adapted from Robert Stoller's *Sex and Gender* published in 1986, by a wide range of (critical) scholars across disciplinary boundaries. Martin and Voorhies title their chapter with *sex* yet talk about *gender*. They are not the only ones who establish the distinction yet conflate the categories.

8. This is also quite surprising, since one may presume that Herdt came across Judith Butler's (1990, 1993) work at one point, not just because they were located not to far from each other in California, but rather because he is deeply involved in thinking about sex and gender. However, the new paper is nevertheless called "Mistaken Sex."

9. Over the years, Herdt translated *kwolu-aatmwol* a number of ways. In 1981 he explains that the Sambia 'identify' with two sexes male (*aatmwul*) and female (*aambelu*). In 1988, he states that beside this 'sexual dimorphism,' Sambia recognize the *kwolu-aatmwol*. The term is a compound morpheme referring to 'male-like-thing' (*kwolu*) and an 'adult person, masculine' (*aatmwul*). This emphasizes the *transformational quality of changing* from a "male-like thing into masculinity" (Herdt & Davidson, 1988: 38). In 1990, Herdt states that *kwolu-aatmwol* indexes "male thing-transforming-into-female-thing" (1990: 439). In 1994 Herdt translates it simply as "changing into a male thing" (1994: 432). Herdt was not able to find an accurate translation of the term *kwolu-aatmwol*. The change of translation in Herdt's course of research reveals the inadequacy of translation.

10. Shame and secrecy as the most dominant feature of intersexualization have recently been tackled by intersex social movements (Preves, 2003). The Internet has played a crucial part in the development of these movements and has helped to connect people who are geographically dispersed. Intersex movements, which have developed since the middle of the 1990s, have now reached great numbers of participants and have shown a considerable growth in membership (e.g., ISNA, AISSGUK, Bodies Like Ours, OII international).

11. Herdt's work on ritualized homosexuality in *Guardians of the Flutes* (Herdt 1981) had great impact on the anthropology of sexuality. Baldwin and Baldwin (1989) for example, based their learning theories of sexuality on the data on the Sambia provided by Stoller and Herdt. They develop a universalized learning theory of homosexual and heterosexual orientation via an uncritical adaptation of Herdt's concept of ritualized homosexuality. Harriet Whitehead argued, however, that those anthropologists who have interpreted institutionalized homosexuality and transgender phenomena in terms of bodily desire or psychological predisposition are missing the point (Whitehead, 1981). She suggests that "a good many of the cross-cultural investigations have been, explicitly or implicitly, aimed at mustering support for one or another interpretation of 'our' homosexuality rather than at laying bare the meaning of 'theirs'" (80). She furthermore argues that "on the one hand 'spontaneous' desire should be considered independently from institutionalized practices and that on the other hand culturally established operations or acts are better understood in terms of the meanings allocated to them in their specific contexts" (80).

Bibliography

Anthropology Today (1998). *News*, *14*(4), 30.

Anthropology Today (2000). *News*, *16*(1), 26.

Arnold, D. (1993). *Colonizing the body. State medicine and epidemic disease in nineteenth-century India*. Berkeley: University of California Press.

Baldwin, J. I., & Baldwin, J. I. (1989). The socialization of homosexuality and heterosexuality in a non-Western society. *Archives of Sexual Behavior*, *18*(1), 113–129.

Bermant, G., & Davidson, J. M. (1974). *Biological bases of sexual behavior*. New York: Harper and Row.

De Cecco, J. P. (1986). Preface. In E. Blackwood (Ed.), *The many faces of homosexuality. Anthropological approaches to homosexual behaviour* (pp. ix–x). New York, London: Harrington Park Press.

Derrida, J. (1982). *Margins of philosophy*. Chicago: Chicago University Press.

Elliston, D. (1995). Erotic anthropology: "Ritualized homosexuality" in Melanesia and beyond. *American Ethnologist*, *22*(4), 848–867.

Fanon, F. (1965). *Studies in a dying colonialism*. New York, London: Earthscan Publications.

Fausto-Sterling, A. (2000). *Sexing the body. Gender politics and the construction of sexuality*. New York: Basic Books.

Forster, P. (1973). Empiricism and imperialism. In T. Asad (Ed.), *Anthropology and the colonial encounter* (pp. 23–38). London: Ithaca.

Foucault, M. (1978). *History of sexuality*. London: Penguin Books.

Foucault, M. (1994). *The order of things: An archaeology of the human sciences*. New York: Vintage Books.

Grosse, P. (2000). *Kolonialismus, Eugenik und bürgerliche Gesellschaft in Deutschland 1850–1918*. Frankfurt am Main, New York: Campus.

Herdt, G. (1981). *Guardians of the flutes: Idioms of masculinity*. New York: McGraw-Hill.

Herdt, G. (1990). Mistaken gender: 5-alpha reductase deficiency and biological reductionism in gender identity reconsidered. *American Anthropologist*, *92*(2), 433–446.

Herdt, G. (1994). Mistaken sex: Culture, biology and the third sex in New Guinea. In G. Herdt (Ed.), *Third sex third gender. Beyond sexual dimorphism in culture and history* (pp. 419–446). New York: Zone Books.

Herdt, G. (2003). *Secrecy and cultural reality*. Ann Arbor: University of Michigan Press.

Herdt, G., & Davidson, J. (1988). The Sambia "Turnim-Man": Sociocultural and clinical aspects of gender formation in male pseudohermaphrodites with 5-alpha reductase deficiency in Papua New Guinea. *Archives of Sexual Behavior*, *17*(1), 33–56.

Herdt, G., & Stoller, R. (1987). Der Einfluss der Supervision auf die ethnographische Praxis. In H. P. Duerr (Ed.), *Die wilde Seele* (pp. 177–199). Frankfurt am Main: Suhrkamp.

Hoad, N. (2000). Arrested development or the queerness of savages: resisting evolutionary narratives of difference. *Postcolonial Studies*, *3*(2), 133–158.

Holmes, M. (2008). *Intersex: A perilous difference*. Selinsgrove, Pennsylvania: Susquehanna University Press.

Imperato-McGinley, J., Peterson, R. E., Gautier, T., & Sturla, E. (1979a). Androgens and the evolution of male-gender identity among male pseudohermaphrodites with 5-alpha reductase deficiency. *New England Journal of Medicine*, *300*(22), 1233–1237.

Imperato-McGinley, J., Peterson, R. E., Stoller, R., & Goodwin, W. E. (1979b). Male pseudohermaphroditism secondary to 17 beta-hydroxysteroid dehydrogenase deficiency: gender role change with puberty. *Journal of Clinical Endocrinology and Metabolism, 49*(3), 391–395.

Jordan-Young, R. M. (2010). *Brain storm: The flaws in the science of sex differences.* Cambridge, MA: Harvard University Press.

Karkazis, K., Jordan-Young, R., Davis, G., & Camporesi, S. (2012). Out of bounds? A critique of the new Policies on hyperandrogenism in elite female athletes. *The American Journal of Bioethics, 12*(7), 3.

Kagan, S., & Hafez, E. S. E. (Eds.). (1981). *Pediatric andrology.* Kerns, Virgina, The Hague: Martinus Nijhoff.

Klöppel, U. (2008). XX0XY ungelöst: Die medizinisch-psychologische Problematisierung uneindeutigen Geschlechts und Trans/Formierung der Kategorie Geschlecht von der Zeit der Aufklärung bis in die Gegenwart (Unpublished doctoral dissertation Universität Potsdam).

Knauft, B. M. (1999). *From primitive to postcolonial in Melanesia and anthropology.* Ann Arbor: University of Michigan Press.

Levinas, E. (1998). *Totality and infinity. An essay on exteriority.* Pittsburgh: Duquesne University Press.

Martin, K., & Voorhies, B. (1975) *Female of the species.* New York: Columbia University Press.

Money, J. (1968). *Sex errors of the body and related syndromes. A guide to counseling children, adolescents and their families.* Baltimore: The Johns Hopkins Press.

Money, J., Hampson, J. G., & Hampson, J. L. (1955). Hermaphroditism: Recommendations concerning assignment of sex, change of sex, and psychologic management. *Bulletin of Johns Hopkins Hospital, 97*(4), 284–300.

Moore, H. L. (2007). *The subject of anthropology. Gender, symbolism and psychoanalysis.* Cambridge: Polity Press.

Oudshoorn, N. (1994). *Beyond the natural body: An archeology of sex hormones.* London, New York: Routledge.

Oudshoorn, N. (2001). On bodies, technologies, and feminisms. In A. Creager, E. Lunbeck, & L. L. Schiebinger, (Eds.), *Feminisms in twentieth-century. Science, technology, and medicine* (pp. 199–213). Chicago: University of Chicago Press.

Preves, S. (2003). *Intersex and Identity. The Contested Self.* London: Rutgers University Press.

Spiro, M. (1993). Is the western conception of the self "peculiar" within the context of the world cultures? *Ethos, 21*(2), 107–153.

Spivak, G. C. (1988). Can the subaltern speak? In C. Nelson and L. Grossberg (eds.), *Marxism and the Interpretation of Culture.* Urbana, Chicago: University of Illinois Press, 271–313.

Stauder, J. (1993). The "relevance" of anthropology to colonialism and imperialism. In S. Harding (Ed.), *The "racial" economy of science. Toward a democratic future* (pp. 408–432). Bloomington, Indianapolis: Indiana University Press.

Stoller, R. (1975). *Sex and gender, Vol. 2: The transsexual experiment.* New York: Jason Aronson.

Strathern, M. (1988). *The gender of the gift. Problems with women and problems with society in Melanesia.* Berkeley, London: University of California Press.

Venn, C. (1999). Occidentalism and its discontents. In P. Cohen (Ed.), *New ethnicities, old racisms?* (pp. 37–62). London, New York: Zed Books.

Whitehead, H. (1981). The bow and the burden strap: A new look at institutionalized homosexuality in native North America. In S. Ortner, & H. Whitehead (Eds.), *Sexual meanings: the cultural construction of gender and sexuality* (pp. 1–28). Cambridge: Cambridge University Press.

Wilson, J. (1999). The role of androgens in male gender role behavior. *Endocrine Reviews, 20*(5), 726–737.

Young, R. (1995). *Colonial desire. Hybridity in culture, theory and race*. London: Routledge

Excursus
Bound to the Third?

Anthropologists have long been fascinated by cultures which recognize more than two genders (see, for instance, Kessler & McKenna, 1978; Malinowski, 1929; Martin & Voorhies, 1975; Mead, 1950). Symbolic organizations which show different, multidimensional societal compositions other than the Western sex-gender-sexuality-system have been used to demonstrate that binary sex_ gender is not a universal and obvious biological fact and, as such, that Western conceptualizations of sex_gender are dependent on a range of disciplinary cultural, symbolic, and structural regimes. Ethnological findings were thought to challenge the supposed universality of the heteronormative and dichotomous order in Western societies. These ethnological endeavors can be read as the attempt to prove or disprove the concordance between sex and gender. In this context a specific notion of the Third, as a different category than the 'first' sex (male) and the 'second' sex (female),[1] has had a revival for mediation of knowledge on sex, gender, and sexuality. It is this configuration of the Third that will be of interest in this excursus. I argue that the various attempts to fill the concept of the Third with a meaning have taken different forms throughout time. In recent anthropological research, however, the Third has been occupied with the process of cross-cultural intersexualization.

The Third as a category has recently been invigorated by a variety of researchers coming from different backgrounds and disciplines carrying different political and/or academic agendas. The Third has specifically re-appeared in ethnological research as an overall term for an array of diverse forms of human experiences and social, cultural, and bodily existence (e.g, Herdt, 1994; Martin & Voorhies, 1975). It has been applied to designate and name the relations between experience and socio-cultural and bodily existence—to interrogate what Westerners understand with the term identity. Several forms of living and expression have been put in the same category as the Third. Introductory anthropology textbooks commonly cite the *hijra* of India, the *berdache*/Two-Spirit of native North America, the *xanith* of the Arabian peninsula, and the female husbands of Western Africa as examples of a third sex_ third gender (Cucchiari, 1981; Herdt, 1994; Ortner, 1981; Roscoe, 1994, 2000). The concept of the Third has been applied with a variety of different meanings and could be described as a *concept in search of a referent*. In this excursus I focus on the history of the *berdache* and the Two-Spirit movement,[2] which re-claimed this history to create something new.

In this excursus, I first highlight anthropological research which uses the *berdache* as an example of a Third. Hereby, I draw on the etymology of the term *berdache* to exemplify the use of an empty signifier that has traveled through centuries and continents to be rested on different bodies and identities. I chose the *berdache* because researchers nowadays see the positions that persons formerly described as *berdache* as being rightly put as a third (or fifth or fourth) place in relation to the Western binary sex_gender system. Interestingly, the descriptions of the *berdache* tend to feature non-normative maleness. This is a recurrent theme of the Third in anthropological research and highly significant in relation to the male bias that can often be found in research into a supposed third sex or third gender. The next paragraph discusses the Two-Spirit movement, which has defeated the foreign (etic/external) term of *berdache*. Here, I introduce the reader to one notion of the Third, which I do regard as useful (in some theoretical settings) to deconstruct the dichotomous universalized sex-gender-sexuality system. Marjorie Garber has elaborated on the option to use a third in the service of irritation that exceeds the dichotomous organization of things and beings. I then briefly touch upon Homi Bhabha's concept of the third space, which he introduced to delineate the move from an 'us-them' dualism to a mutual sense of 'both/and.' By drawing on him, I argue that the Two-Spirit movement can undermine the colonizing move of anthropological accounts of the *berdache*.

The *Berdache*

For centuries, the term *berdache* has been used by European colonizers to describe people in Native North American cultures who they perceived as different in their role, expressions, and lives in regards to the culture they were investigating, yet, more importantly, as different to their known culture.[3] Historian Rudi Bleys, in his *Geography of Perversion,* argues that ethnographic information on the *berdache* possibly influenced the debate on hermaphrodites at the beginning of the twentieth century and therefore changed conceptualizations of sexuality and sexual identity substantially (1995: 71). I suggest that the *berdache*, as a Third, was indeed fundamental for cross-cultural intersexualization. The term *berdache* has an interesting etymology. Narratives on this etymology differ but it seems that this word was first used as *bardash* by the first French colonizers in New France (Canada) and that it originates from the Arabic word *bardaj*, which means 'slave' (Marquette, 1900 [1674]: n. 26 in Hultkrantz, 1983: 459). Angelino & Shedd note in 1955 in their extensively cited paper that:

> an etymological investigation of the English word "berdache," or "berdash," indicates that it derived from the French word "bardash," which derived from the Italian word "berdascia," which derived from the Arabic word "bardaj," which derived from the Persian "barah." While the

word underwent considerable change the meaning in each instance remained constant, being a "kept boy," a "male prostitute," a "catamite."[4]

(121)

They furthermore state that there "was no generally accepted concept of berdache" (121). However, in most of the early accounts the term was used by colonizers to describe 'acts of sodomy' and 'transvestism' in North American Native cultures from the beginning of the eighnteenth century (e.g., Bossu 1768 in Angelino & Shedd, 1955: 122). Jonathan Katz, historian of sexuality, records even earlier uses in the sixteenth century (1976: 285–286). Most of the accounts use the term when they either encounter a person inhabiting a specific social role or a specific choice of erotic object. Mostly explorers and missionaries used the term to describe the supposedly "perverted gender behaviors" and "unnatural sexual practices" (1976: 288–291). To anthropologists the term *berdache* has in Angelino's and Shedd's time become synonymous with "transvestism and effeminacy" (1955: 122).

Charles Callender and Lee Kochems composed a comprehensive collection of research into the *berdache* which they called "The North American Berdache" (1983). They surprisingly do not reflect on the use of the term at all but list 113 cultures, on the basis of an analytical synthesis of published and unpublished accounts, which recognize the 'berdache status' although none of them, of course, uses this term. Their extensive discussion, however, shows that there are no common features to be nailed down and that not just every culture but every person expressed his_her identity differently and was differently perceived by their people. One would assume that the heterogeneity of the cultures in North America, however, seems likely to prevent any kind of overall description and lumping together under one term that is not even *emic* in its origin. Any *etic* term would also reflect some of the Euro-American biases that the researcher brings to the field (see Epple, 1998: 268). The neologisms *emic* and *etic* were derived from an analogy with the terms 'phonemic' and 'phonetic.' They were coined by the linguistic anthropologist Kenneth Pike (1962) who drew the parallel between the two perspectives that can be employed in the anthropological study of a society's cultural system, with the two perspectives that can be used in the linguistic study of a language's sound system. *Emic* in this framework means a description coming from within the culture, *etic* is a description of a behavior or belief by an observer, which can be applied to other cultures and is therefore regarded as 'culturally neutral.' The *emic/etic* dichotomy leads to issues about the very nature of objectivity and therefore provoked some controversy for example between the cultural anthropologists Marvin Harris (1968) and Kenneth Pike. Pike believed that all claims to knowledge are ultimately subjective; hence, objectivity is impossible for him. Harris did believe in objective knowledge production and deemed that for anthropology it is necessary to aspire to gain such knowledge in order to be taken seriously as an academic discipline (see also Headland, Pike, & Harris, 1990). However, in Callender's and Kochem's account this discussion seems to be absent.

One of the early articles on the *nádleehí*, who are subsumed under the umbrella term *berdache* by Callender and Kochems, which was also extensively cited since its publication in 1935, is that by anthropologist W. W. Hill who did research in Navaho culture. Hill begins his article by stating that "unlike our own society, many primitive societies recognize in a social sense, and include in their culture pattern a place for those people whose psychic or physiological peculiarities set them apart from the normal" (1935: 273). Hill reports on the status of the *nádleehí* in Navaho cultures and states that "the concept of the nádleehí is well formulated and his *[sic!]* cultural role well substantiated in the mythology" (273). The term *nádleehí* according to Hill's account has been used for 'hermaphrodites' and 'transvestites,' whereas the hermaphrodites are considered to be the real *nádleehí*. Moreover, the *nádleehí* seems to be conceptualized as male in Hill's work, which is an interesting feature also in intersexualization as I have demonstrated in Chapter 5 in this volume. In Hill's account the *nádleehí* he interviewed supposedly does not fit in the special role, which is provided for her_him by the culture. The personal pronouns and the language to address kin and clan members used by the *nádleehí* Hill interviewed did not have one referent but drew interchangeably on both sexes_genders. Hill concludes that his interviewee "has failed to make the personal adjustment which her culture makes possible" which for him leads to the assumption that this "is probably also true of others" (279). Hill's conclusion mirrors his inability to imagine that this might be not conceptualized as failure by the Navaho culture but as a way of expression of the *nádleehí* position. His report cannot handle the fact that the *nádleehí* might enable multiple and singular ways to handle the cultural position which is offered by the Navaho culture. The contradictory information he gets from his interviewees leads him to assume that single people cannot adjust to a societal position which combines aspects of the other positions without having to decide or be unequivocally adapted to either or the Third. His narrative cannot accommodate the semantic and symbolic difference, which seems to be part of the *nádleehí* subject position. Subsequently, Hill's interpretation cannot but fail to describe appropriately from an outside position, which is located in a binary sex_gender system. In 1955, at the same time in which Money (see Chapter 1 in this volume) started writing about sex and gender as distinct, Angelino and Shedd review Hill's account on the *nádleehí* in the Navaho culture. They state that Hill reported that there are "those who pretend to be nádleehí who are, according to them transvestites who should be distinguished from the 'real nádleehí' who are hermaphrodites" (1955: 125). They conclude from Hill's narrative that in Navaho cultures there "is a definite differentiation between those individuals who are as they are because of obvious physiology and those who are as they are because of psychology" (125). From Hill's writing this distinction does not become clear. However, Angelino and Shedd imply that the Navaho make a distinction between real and unreal *nádleehí* on the grounds of physiology. They conclude that "intersexed individuals are special cases and consideration of them from a scientific point of view as *berdache* is to lose sight of the *fundamentum divisionis*" (124). Angelino and Shedd consequently create recommendations of how to write of *berdache* and about who is to be included in the rubric:

> In view of the data we propose that *berdache* be characterized as an individual of a definite physiological sex (male or female) who assumes the role and status of the opposite sex, and who is viewed by the community as being of one sex physiologically but as having assumed the role and status of the opposite sex. If erotic object is to be noted it should be so designated by the appropriate adjective: heterosexual, homosexual, bisexual, etc. [...]. While a *berdache* is a transvestite, a transvestite is not necessarily a *berdache*. In no instance is a hermaphrodite a berdache. It appears that hermaphroditism is an adequate characterization.
>
> (125)

They here articulate the confusion and helplessness of Western scholars to grasp and describe the *berdache*. However, they seem quite confident in using sexological terms, which are related to the discourse that has just been established in the USA. This becomes evident when these sexological terms are used to delineate behavior, desires, and roles. They do not yet use the terms of sex and gender; yet, their account exemplifies the distinction between biology and culture. Nevertheless, they want to give clear guidelines of how to handle this confusion, which are to be followed if one wants to arrive at a general or universal use of the term *berdache*. The ordering system they propose is bio-medical; a 'definite physiological sex' is presumed according to which the 'opposite sex' has to be measured and designated. The ranking of 'erotic object choice' is also dichotomously and irreversibly fixed as either homo-hetero-or-bisexual. The note on hermaphroditism shows that the *berdache* has to be clearly delineated from the pathologizing notions of Western bio-medicine. Angelino and Shedd state that "in no instance is a hermaphrodite a *berdache*" (125). Therefore, they want to understand the *berdache* in simple binary terms—otherwise the formerly mentioned criteria for physiological sex would not apply anymore. Yet, they add that "it appears that hermaphroditism is an adequate characterization" (125) leaving the reader confused as to *what* hermaphroditism is adequate. The hermaphrodite cannot take a position in this system because the notions of same and opposite were to be reviewed. What does become clear here is the desire to fill an empty signifier with a clear set of meanings, which obscure the particularities.

Throughout time, the *berdache* took on several different roles and positions in regards to Western biological and psychological criteria—none of them could unify the researchers. However, research into the *berdache* (and other Thirds) represents a projection of a Western fantasy of an 'exotic' symbolic order and other and has been employed for political ends in regards to Western homophobia and transphobia. Cultural anthropologist Nico Besnier states that an identification with *berdache* figures is "understandable in the context of lesbians' and gays' struggle for a political voice in postindustrial societies" and also the desire to use these figures to "demonstrate that preindustrial societies are more 'tolerant'... or 'accommodating' of erotic diversity and gender variation than 'the West'" (1994: 316).

Morgan Holmes detects that current research about the *berdache*, namely research by historian Will Roscoe (1994), is driven by gay activist and masculinist agendas (Holmes, 2004). Holmes describes this as an ongoing exploitation of the colonized for political purposes, in this case for the enhancement and support of the gay movements in North America in the 1980s and 1990s. Moreover, the tendency to employ and direct the Third towards a consolidation of (here, gay) masculinity is a recurrent trope (see below). Holmes states that "if more recent anthropological work avoids the early finger-pointing behaviours of colonizers and theologically minded scholars, that does not mean that it avoids the pitfalls of using 'others' to further its own domestic agendas" (2004: 5). Holmes furthermore traces the history of ethnological research and shows that it is tightly bound to missionary endeavors and the colonial order. In general, whenever the *berdache* is talked about anthropologists assume and base their argument on the Western coordinates of a male and female without explicit deliberation or reflection on indigenous constructions of bodies and sexualities (see Goulet, 1996: 685). Anthropologist Kath Weston (1993) criticizes the term *berdache* as another catch-all-term, which describes and therefore limits this identity position to a position that is again constrained by Western binary terms. Carolyn Epple, also a feminist anthropologist explored the narratives about the First Nation North American category of the *nádleehí* and writes that:

> Ironically, casting them [the people subsumed under the term *berdache*] as such does not subvert but reifies—indeed is based upon—the very system it is intended to dismantle: the binary gender system and its assumed natural coherence among sex, gender, and desire. In setting up nadleehi (and presumably others) as belonging to a "third (or fourth, fifth and so on) gender," theorists reify Man and Woman as binary opposites, using them as standards by which to identify "alternates."
>
> (1998: 273)

Epple here supports my main argument when it comes to intersexualization. The quest for a scientifically verifiable distinction between male and female is based upon creating a Third in order to reify the binary. Moreover, as Epple analyzes it, "the term *alternate gender* suggests that in mixed-gender behaviors there is evidence for an altogether different gender" (1998: 267). In fact, any of these terms that are based on notions such as *alternate, ambiguous, either/or, neither/nor* and so on are problematic because they imply the status of a hybrid constellation for the non-normative and in the same vein install the non-hybrid or pure, real, and normal character of the two sexes_genders. This kind of anthropological research into third sexes_genders, therefore, wishes to fully understand and describe these complex structures and relationships even though the object is not subsumable under any *one* single category. In their political endeavors to reference other cultures as an anti-homophobic strategy they produce a tautological circle that essentializes and produces othering processes. Evan Towle and Lynn Morgan (2002) in their essay on "Romancing the Transgender Native, Rethinking the Use of the 'Third Gender' Concept" state that the third gender concept is "by nature flawed because it subsumes all non-Western, nonbinary identities, practices, terminologies, and histories. Thus it becomes a junk drawer into which a great non-Western gender miscellany is carelessly dumped" (484).[5]

Moreover, as philosopher of science Myra Hird argues, "replacing a two-sex model with a 10-sex (or 20 or 30) model does not in itself secure the abolition of gender discrimination, only perhaps that the mental gymnastics required to justify such discrimination becomes more complex" (2000: 358). The supposed discovery of third, fourth, or even more genders reifies the binary opposition of man and woman rather than disrupting it, and hence imposes sex(ualiz)ed and gendered constructs that may be inapplicable. The assumed rigid, preset meanings for masculine and feminine confer 'identity' on a person on the basis of only a few traits and, as such, disregard the complexity of any culture and all subject(ivitie)s. The question remains if an adaptation of the disciplining forces in the production of Western knowledge about identity and embodiment can foster liberation from essentialist and limiting notions. Or does this construction of a third identity position on the grounds of a third morphological category have the effect of making non-heteronormative and non-binary positions only available for specifically medically identified/diagnosed/categorized persons and therefore pathologizes these people? Epple's critique centers on the different uses of Western terms for other people's sexuality and gender conceptualizations. Epple states that future research:

> should begin with attention to the history of specific cultures, the exploration of multiple systems of meanings (as in other interpretations of Sa'gh Naaghai Bik'eh Hozhb and nadleehi), and the identification of culturally specific and relevant constructs. Such a particularistic focus does not preclude cross-cultural research; indeed, it should enhance intercultural comparison by ensuring that research proceeds from culturally valid classifications. With locally salient meanings finally reinserted, new ways to organize the discourse can emerge, ways that take the analysis beyond gender and sexual practices and redefine the discourse itself.
>
> (1998: 280)

Kath Weston contests the move to employ indigenous categories because she thinks them "as no more neutral in its effects" (1998: 159). According to Weston, using indigenous terms is intrinsically othering because they construct the subject of inquiry as always and already other. She argues that these 'foreign terms' become complicit in a form of Orientalism in which language simply reifies difference and buttresses ethnographic authority. In my eyes this always comes down to the hierarchies that are cemented with the reference to difference. Of course, it is the ethnographic authority which has to be handled very carefully but the question is if there are any terms in the Western anthropologist's toolbox which do not do the job of othering. Some ethnologists recognized that there are conceptual limitations in the approach to third sexes_genders and that any interpretation is insufficient because of the inadequacy of social and linguistic concepts. Will Roscoe acknowledges that what people write about third sex_gender categories "reflects more the influence of existing Western discourses on gender, sexuality and the Other than on what observers actually witnessed" (1994: 330). Therefore, with the (ab)uses of the category of the Third, the problem of generalization of the huge range of immensely diverse cultural categories which have been explored by Western scholars emerges. However, some might not be able to identify as a Third as they consider themselves as totally distinct from the parameters applied to man and woman and, therefore, also from the parameters applied to the Third as they are derivates from the formerly mentioned (Towle & Morgan, 2002). The construction of the Third remains in the Western notion of identity as stable through life and therefore limits again the possibility of identification beyond Western notions of personhood. Being bound to the Third as a morphological, psychological, or identificatory category disciplines the perception of the huge variety of identities, practices, and processes according to the coordinates of a binary and heteronormative sex-gender-sexuality-system. Moreover, being bound to the Third colonizes the symbolic system of others culture while silencing that there are also intersexualized and transgendered people in the (kn)own culture.

The Two-Spirit Movement

The term two-spirit was first coined in 1990 at a gathering of Native Queer/Two-Spirit people in Winnipeg, Canada, as a means to replace the word *berdache*. It is a contemporary term that was adopted from the Northern Algonquin word *niizh manitoag*, meaning two spirits. It is meant to signify the embodiment of both feminine and masculine spirits within one person (Anguksuar, 1997)[6] and therefore signifies a different conception to the Western one of sex and gender. To Terry Tafoya, two-spirit is a verb, whereas gay is a noun which one has to read as a metaphoric statement and which is used in the context of a noun signifying a place, person, or thing, and a verb being associated with processes, actions, and interactions (Tafoya, 1992: 256; see also Epple, 1998). Qwo-li Driskill, a Cherokee Two-Spirit himself,[7] describes two-spirit as "also a way to talk about our sexualities and genders from within tribal contexts in English" (Driskill, 2004: 52). Driskill, in the article "Stolen From Our Bodies. First Nations Two-Spirits/Queers and the Journey to a Sovereign Erotic" (2004) writes about First Nations people and embodiment, desires, and practices and states that these are:

> braided with the legacy of historical trauma and the ongoing process of decolonization. Two-Spirits are integral to this struggle: my own resistance to colonization as a Cherokee Two-Spirit is intimately connected to my

continuing efforts to heal from sexual assault and the manifestations of an oppressive overculture on my erotic life. Like other Two-Spirit people, I am making a journey to a Sovereign Erotic that mends our lives and communities.

(51)

Therefore, the term two-spirit is not a historical term, but rather a neologism that is supposed to resist colonial definitions and to be an alternative to these. Driskill argues that the term is determined to express "our [First Nation people's] sexual and gender identities as sovereign from those of white GLBT movements" and, furthermore, cautions that "the coinage of the word was never meant to create a monolithic understanding of the array of Native traditions regarding what dominant European and Euro-American traditions call 'alternative' genders and sexualities" (51). Jacobs & Thomas state that the term is "a contested compromise to move forward the debate in eliminating culturally inappropriate terms," and that it is designed to include a wide range of Native people such as: "cross-dressers, transvestites, lesbian, gay, transgendered, or [those] otherwise 'marked' as 'alternatively gendered' within tribes, bands, and nations where multiple gender concepts occur" (1994: 7). However, there seems to be a difference between urban First Nation people and those who live on the reservations. Sabine Lang mentions that it is mainly urban contemporary Native American lesbian and gays who started to recover the history of "gender variant/alternate" people of the pre-reservation days as their predecessors. "Two-spirited or two-spirit," Lang writes, "is the term that is used to refer to themselves to express that continuum" (1999: 91).

The terms 'gender variant' or 'alternate' are the new anthropological terms applied in research. Lang uses them although she realizes that it is problematic in terms of the implication of biological or stable sex. It was explained to Carolyn Epple by one of her Navajo cultural teachers that Navajo people "usually did not talk about genitalia; thus, who was and was not a 'true' hermaphrodite may not have been shared beyond immediate family members" (1998: 271). Lang's research, moreover, is problematic because she uses the terms 'man-woman' and 'woman-man' to describe the *nádleehí* people she researched and interviewed. In this manner she falls back into an inadequate, or at least simplifying, scheme of naming something that might not be possible to be named in a different linguistic and symbolic framework. Robert Padgug describes this as an "enshrinement of contemporary sexual categories as universal, static, and permanent, suitable for the analysis of all human beings and all societies" (1979: 8). For Padgug, different societies:

share general sexual forms [which does] not make the contents and meaning of these impulses and forms identical or undifferentiated. They must be carefully distinguished and separately understood, since their inner structures and social meanings and articulations are very different.

(1)

Classificatory processes are based on simplification and are furthermore hindered by the possibility of translating per se and of translating complexity. One of Epple's interviewees states that:

> So with nádleehí, like the clothing and stuff, that is so artificial, so why make a big stink about it? If you were to look at that person, at all the natural processes [such as the air, sun, etc.] interconnecting to him, that alone would fill up books and books. Then you get to this one part, this artificial part about his clothing. In a drawing of him with all of his interconnections, you'd have to magnify that artificial part a million times even to see it.
>
> (1998: 278)

Any attempt to translate or describe would fail because the complexity and the interconnectedness of lives, people, and their culture can only be flattened and therefore misrepresented. It can never be complete. The two-spirit person (here in Navajo culture) is according to Epple's interviewee seen as "the unique configuration of all natural processes coming into her or him," not as a handful of traits (1998: 278). Epple continues to quote her interviewee on his views and he states: "one exists as both, male and female aspects hold as true for nádleehí as it does for a mountain, a tree, a woman, or a man." He goes on: "Everything is two, so how can you have this as a third? You don't have man, woman, and another" (278). Again, this is interesting to position side by side with Driskill's statement who says "We simply *are*" (Driskill, 2004: 55 [emphasis in original]). Two-spirited people, as the only constant feature of the research on them shows, act and interact to define and redefine themselves, especially in a neo_colonial era. Tafoya describes this dynamic ongoing and shifting process of self-definitions as based on 'Native tradition' that "emphasizes transformation and change, and the idea that an individual is expected to go through many changes in a lifetime" (1992: 257). So the term two-spirit seems to be fluid and dependent on situations, interpretations and geo-political positioning by people who somehow identify with First Nation traditions but are wary of constant change, not just in these socio-cultural settings but who also have differing self-conceptualizations.

Different writers who, in some way or the other, identify with the term two-spirit state different possibilities of describing and using the term. To Driskill's "knowledge as a non-fluent Cherokee speaker, there is currently no term in Cherokee to describe Two-Spirit people" (2004: 55). Driskill explains the complexity and the choice of the term two-spirit accordingly in a footnote: Driskill uses it "because it does not make me splinter off sexuality from race, gender from culture. It was created specifically to hold, not diminish or erase, complexities. It is a sovereign term in the invaders' tongue" (63, n.3). Therefore, for Driskill, the process of translating two-spiritness into "white communities [symbolic and linguistic frameworks] becomes very complex" (52). Therefore, there are many obstacles to translating the differences in symbolic systems. Differing conceptualizations of embodiment, desires, and practices are either lost, only partially transferred, or obscured by colonial processes. However, as a self-empowering term and one that marks distinctions to the governing white/Western mainstream it seems to prove valuable and has also been used powerfully by First Nation people to, on the one hand draw on history, but on the other to emphasize that this history (as all history) is a work in progress. The term two-spirit seems so flexible and under constant reevaluation and redefinition that it might eventually be a historically correct term to describe the present. It might hold the promise of designating an identity which is composed of multiple sites, existing as a political term in the first place and empowering people while discovering a lost narrative, which they can apply to new narratives. Kimberley Balsam et al., quote Raven Heavy Runner, Two-Spirit, Blackfeet, stating that:

> this two-spirit movement is of re-establishing our culture. The two-spirit movement if anything is a decolonization process, to support the Native community and to reclaim those roles we used to have. We're doing this not for ourselves, but for those who can't—those who are young and just coming out, and for elders who haven't felt supported throughout their life.
>
> (2004: 287)

In this quote it is obvious that the two-spirit movements aim at a decolonization process. Additionally, it shows that the Two-Spirit movement combines the political agenda to support different generations and thereby aims at bringing together past and future. The Two-Spirit movement could be described as being located in three third spaces; in-between generations, in-between genders, and in-between colonization and de-colonization. These thirds can be made productive for something new.

The Third Space

Marjorie Garber in her book *Vested Interests* on cross-dressing argues that "thirds are analytically useful because they upset the binary and encourage flexibility" (1992: 10). For Garber cross-dressing and bisexuality have the potential to disrupt "easy notions" of binarism and of questioning the "categories of 'female' and 'male, whether they are considered essential or constructed, biological or cultural" (10).[8] And she elaborates:

> the "third term" is *not* a *term*. Much less is it a *sex*, certainly not an instantiated "blurred" sex as signified by a term like "androgyne" or "hermaphrodite", although these words have culturally specific significance at certain historical moments. The "third" is a mode of articulation, a way of describing a space of possibility. Three puts in question the idea of one: of identity, self sufficiency, self-knowledge.
>
> (11)

Garber rejects the idea that the Third is principally a word, sex, or specific referent of any kind. Garber is particularly interested in the ability of multiple kinds of 'thirds' to disrupt multiple binary categories and symmetries by placing them in larger, messier contexts. To her, it is the task to put the Third into a contextualization which takes it out of the dual relation in which it formerly stood and places it in a larger chain. The Third to her could serve to reconfigure the relationship of the original dichotomy and to question the "identities previously conceived as stable, unchallengeable, grounded, and 'known'" (13). Here, the Third has the potential to question binary thinking and to introduce crises.[9]

Homi Bhabha talks about a third space through which he rethinks assumptions about culture and identity from an 'us-them' dualism to a mutual sense of 'both/and.' I argue that a creation of such a third space can be detected in the two-spirit movement. The third space in Bhabha's notion is a place where identity is negotiated and where people become neither this nor that but their unique subjectivity (English, 2005). Furthermore, the third space is used to signify the place where this negotiation happens and where identity is constructed and re-constructed. It is also a space where life in its ambiguity, complexity, and hybridity is lived and redefined. The notion of the third space opens up potentials for the creation of new structures of authority and for new interpretations of identity as interdependent, temporary, and fluid. Bhabha cautions us that:

> the construction of the colonial subject in discourse, and the exercise of colonial power through discourse, demands an articulation of forms of difference—racial and sexual. Such an articulation becomes crucial if it is held that the body is always simultaneously inscribed in both the economy of pleasure and desire and the economy of discourses, domination and power.
>
> (1997: 38)

To be able to counter the articulation of this kind of difference, Bhabha developed the concept of hybridity. He uses it to describe the construction of culture and identity within conditions of colonial antagonism and inequity (1994, 1996). Bhabha argues that new hybrid identities or subject-positions emerge from the interweaving of elements of the colonizer and the colonized. To Bhabha, hybridity is the process by which the colonizers try to translate the identity of the colonized (the *Other*) within a singular universal framework, but fail to produce something familiar and in this failure produce something new—the hybrid (Papastergiadis, 1997). Therefore the hybrid is a representation of cultural differences, which are positioned in-between the colonizer and the colonized. For Bhabha, it is the undefined and indefinite *spaces in-between subject*-positions and identities, which become the space of disruption and displacement of hegemonic colonial narratives (1994, 1996). He conceptualizes hybridity as such a form of liminal or in-between space—the third space. Bhabha is aware of the dangers of fixity and fetishism of identities within binary colonial thinking, arguing that "all forms of culture are continually in a process of hybridity" (Rutherford, 1990: 211). However, the history of the concept of hybridity makes its employment problematic because hybridity is historically a violent term describing people who are 'mixed-breeds,' so-called products of 'miscegenation.' It is deeply intertwined with nineteenth-century eugenicist and scientific-racist discourse (Mitchell, 1997; Werbner & Modood, 1997; Young, 1995). Therefore, I do see problems with using it because the references it combines reaffirm the two entities, which are supposedly more 'pure' and then merge into the hybrid. The notion of the hybrid in itself implies that there is such a thing as purity, whether it is in relation to identity, gender, body, nationality, ethnicity, or origin in general. Moreover, whenever there is a third (no matter if it is an identity, a body, or a space) the first two are implicit and are thought of and listed according to a specific hierarchy. This hierarchy makes one of the two entities the more dominant, hegemonically more justified, debatably more important part and the second one less influential, less important, and more likely to be disregarded. The Third is always based on a first and a second and these are not free from their hierarchical reference system. Theorizing the hybrid means to recognize that there is no such thing in the first place because everything is hybrid—is difference—difference and composition.

Yet, negative terms do have an emancipatory potential. I think here of queer or gay, the appropriation of which has powerfully undermined the hegemonic discourse on pathologization (Butler, 1993). Papastergiadis also asks if we should "use only words with a pure and inoffensive history, or should we challenge essentialist models of identity by taking on and then subverting their own vocabulary" (1997: 258). This would be one strategy; however, it only works if there is resonance in hegemonic culture and if there are discourses which can be called upon to counter discrimination. In the case of the two-spirit movement I suggest that, with the use of a term that has no history in a discriminating colonial framework, the appropriation can work better. The two-spirit people can willingly occupy a doubly hybrid position, namely of that between the normative construction of the sexes_genders and that between the colonial and the colonized. The movement emphasizes different subject positions and is wary of not being subsumed under one single notion. Moreover, their appropriation of terms and symbolic references enables many speaking positions in this *third space*.

Bhabha's *third space* is construed as intrinsically critical of essentialist positions of identity and a conceptualization of 'original or originary culture.' Where the colonizers impose a hegemonic and therefore normalizing practice, the strategy of the hybrid opens up a *third space* of/for rearticulation of negotiation and meaning (1996). The *third space* becomes articulation; it becomes a productive and not merely reflective space that engenders new *possibilities*. It becomes a production that disrupts established categorizations of culture and identity. Hybridity can therefore be seen as the antidote to essentialism. Bhabha's concept of hybridity opens up new anti-essentialist possibilities of conceptualizing globalized identities. If it is applied as a working term in its 'purest' sense, meaning that it is not supposed to denote something that is but something in becoming, the Third can work in a third space to disrupt binaries and hierarchies. Therefore, a symbolically working *third* can work as a disruption concerning the epistemic violence (Spivak, 1988), which has been placed upon Native American cultures by the term *berdache*.

Conclusion

Whereas researchers from the West suggest the flexibility of sexuality, in the same move they invent new accompanying categories such as the *berdache* to produce the same empirical findings time and time again. These findings do not challenge the Western categorization of sexuality but reiterate its parameters by positioning the sexuality of the 'other' as 'other.' The gaze upon the mind and body of the colonized subject is articulated in the light of the hierarchy of a neo_ colonial setting. The societal organizations of non-Western societies are placed together under one rubric—the notion of the Third becomes ontologically universalized. In particular, the addition of the Third (as either third sex and/or third gender) reinforces the categories of male and female as natural, normal, and also cross-culturally universal. It is not just the invention of the position of the Third as the third and not, say, the first, and the limitation of the possibly fourth and fifth (and so on) genders_sexes to this single category. Rather, I would argue, the move to draw on bio-medical categories to argue for the historical and social validity of this identity position and the inherent essentializations, which appear in this discourse, are the most problematic. My contention, therefore, is that by juxtaposing all these different accounts, which interrogate different issues in different spaces and different times, the notion of similarity (sameness) rather than difference is evoked. By creating the Third, anthropologists therefore attempt to fill that created 'space' with a positive connotation. Yet, the reference system they hereby draw upon only reproduces the dimorphic model of sex and the dualism of gender identity development. The problem inherent is a basic one: while arguing that there are 'third categories,' the dichotomous sex(ual) difference between male and female is not tackled (neither in biological terms nor in cultural/social ones). However, I see a potential in movements such as the Two-Spirit movement to defeat and challenge essentialist notions as well as the colonizing impetus in anthropology. Different positionalities and the emphasis on difference and composition can provide discourses that produce counter-narratives especially in post-colonial settings, where people re-claim their history and from this create new narratives and futures.

Notes

1. This observation of the first sex as male and the second as female refers to Simone de Beauvoir's *Le deuxième sexe* (1949). From a feminist perspective this hierarchical organization of the sexes_genders is already suspicious because it mirrors the patriarchal structure of Western societies.
2. I use capitals in Two-Spirit, whenever I designate the movement or the self-identification of a person. In other situation where I explain the term I use it without capitals in line with Terry Tafoya who states that two-spirit is a verb. I think this explanation is apt since I see the emphasis in two-spirit in doing and not in being.

3. Research on the *berdache* has never been a unified field. Some claim that the *berdache* as an institution died out in 1698 (Hauser, 1990). Others see this institution revived in current contexts (Roscoe, 1994; Trexler, 2002) or caution against a subordination of the "native American cosmology to the empire of gender" (Murray 1994). *Berdaches*, moreover, came to be spiritual ancestors for gay activists of the present day. In the following, I focus on accounts which have been quoted frequently by scholars who published on the *berdache*, either because they are the oldest and most established accounts (Angelino & Shedd, 1955; Hill, 1935) or because they are the most comprehensive ones (Callender & Kochems, 1983).

4. The etymology of the term catamite is derived from the Latin catamitus and from the Etruscan Catmite. Catamitus is also known as Ganymede; Greek Ganymēdēs who was seduced by Zeus and became his lover. The term catamite in the history of the ancient world designates the often much younger lover in a male couple.

5. It has to be added here that in Towle and Morgan's account the notion emerges that there is a miscellany in non-Western societies, yet, in Western societies, there is uniformity or sameness.

6. Anguksuar is also known as Richard La Fortune or 'Little Man' and organizes International Two Spirit Gatherings.

7. Driskill describes hirself as Queer poet/activist/educator.

8. Garber goes even further to suggest that, far from being a third sexual identity, bisexuality is a sexuality that "puts into question the very concept of sexual identity in the first place" (1992: 15). Bisexuality, proclaims Garber, is a "sexuality that threatens and challenges the easy binarities of straight and gay" (65). In order to make this claim, however, Garber remains reliant upon the very opposition which underpins that of hetero/homosexuality: (sexual) identity versus (fluid) difference. The only difference is that the hierarchical relationship between the two terms is reversed: difference (which in Garber's model is fluid bisexuality) now elevated at the expense of identity (hetero/homosexuality) (see also Angelides, 2001: 4).

9. In Garber's account the transvestite becomes convincingly the Lacanian "intervening term, 'to seem in' "—"the Signification of the Phallus" where the other two either have the phallus or are the phallus. The transvestite as the figure who incorporates the "seeming" (or "appearing" as Garber phrases it) is the substitute for 'having' and for the protection against the threat of loss. And this specific representation only functions in the "psychic economy in which *all* positions are fantasies" (1992: 356).

Bibliography

Angelides, S. (2001). *A history of bisexuality*. Chicago, London: The University of Chicago Press.

Angelino, H., & Shedd, C. L. (1955). A note on berdache. *American Anthropologist*, *57*(1), 121–126.

Anguksuar, L. R. (1997). A postcolonial perspective on Western [mis]conceptions of the cosmos and the restoration of indigenous taxonomies. In S.-E. Jacobs, W. Thomas, & S. Lang (Eds.), *Two-Spirit People: Native American Gender Identity, Sexuality and Spirituality* (pp. 217–222). Urbana: University of Illinois Press.

Balsam, K., Huang, B., Fieland, K., Simoni, J., & Walters, K. L. (2004). Culture, trauma, and wellness: A comparison of heterosexual and lesbian, gay, bisexual, and two-spirit Native Americans. *Cultural Diversity and Ethnic Minority Psychology*, *10*(3), 287–301.

Beauvoir, S. de (1949). *Le deuxième sexe*. Paris: Gallimard.

Besnier, N. (1994). Polynesian gender liminality through time and space. In G. Herdt (Ed.), *Third sex third gender. Beyond sexual dimorphism in culture and history* (pp. 285–328). New York: Zone Books.

Bhabha, H. (1994). Frontlines/borderposts. In A. Bammer (Ed.), *Displacements: cultural identities in question* (pp. 269–272). Bloomington: Indiana University Press.

Bhabha, H. (1996). Cultures in between. In S. Hall and P. Du Gay (Eds.), *Questions of cultural identity* (pp. 53–60). London: Sage Publications.

Bhabha, H. (1997). The other question. In P. Mongia, (Ed.), *Contemporary postcolonial theory. A reader* (pp. 37–54). London: Arnold.

Bleys, R. (1995). *The geography of perversion. Male-to-male sexual behaviour outside the West and the ethnographic imagination, 1750–1918*. New York: New York University Press.

Butler, J. (1993). *Bodies that matter: on the discursive limits of 'sex.'* New York: Routledge.

Callender, C., & Kochems, L. M. (1983). The North American berdache. *Current Anthropology, 24*(4), 443–470.

Cucchiari, S. (1981). The gender revolution and the transition from bisexual horde to patrilocal band: The origins of gender hierarchy. In S. B. Ortner, & H. Whitehead (Eds.), *Sexual meanings: the cultural construction of gender and sexuality* (pp. 31–79). Cambridge: Cambridge University Press.

Driskill, Q. (2004). Stolen from our bodies. First Nations Two-Spirits/queers and the journey to a sovereign erotic. *Sail, 16*(2), 50–64.

Epple, C. (1998). Coming to terms with Navajo nadleehi: a critique of berdache, "gay," "alternate gender," and "two-spirit." *American Ethnologist, 25*(2), 267–290.

English, L. (2005). Third-space practitioners: women educating for justice in the global south. *Adult Education Quarterly, 55*(2), 85–100.

Garber, M. (1992). *Vested interests. Cross-dressing and cultural anxiety*. New York: Routledge.

Goulet, J.-G. (1996). The "berdache"/"Two-Spirit": A comparison of anthropological and native constructions of gendered identities among the Northern Athapaskans. *The Journal of the Royal Anthroplogical Institute, 2*(4), 683–701.

Harris, M. (1968). *The rise of anthropological theory*. London: Routledge and Kegan Paul.

Hauser, R. (1990). The berdache and the Illinois Indian Tribe during the last half of the seventeenth century. *Ethnohistory, 37*(1), 45–65.

Headland, T. N., Pike, K. L., & Harris, M. (Eds.). (1990). *Emics and etics: The insider/ outsider debate*. Newbury Park: Sage.

Herdt, G. (1994). *Third sex third gender. Beyond sexual dimorphism in culture and history*, New York: Zone Books.

Hill, W. W. (1935). The status of the hermaphrodite and transvestite in Navaho culture. *American Anthropologist, 37*(2), 273–279.

Hird, M. J. (2000). Gender's nature: Intersexuality, transsexualism and the 'sex'/'gender' binary. *Feminist Theory, 1*(3), 347–364.

Holmes, M. (2004). Locating third sexes. *Transformations*. Retrieved February 17, 2010 from www.transformationsjournal.org/journal/issue_08/article_03.shtml.

Hultkrantz, A. (1983). *The study of American Indian religions*. Chico, California: Scholars Press.

Jacobs, S.-E., & Thomas, W. (1994). Native American Two-Spirits. *Anthropology Newsletter, 35*(8), 7.

Katz, J. (1976). *Gay American history: Lesbians and gay men in the U.S.A.* New York: Crowell.

Kessler, S., & McKenna, W. (1978). *Gender: An ethnomethodological approach*. London: The University of Chicago Press.

Lang, S. (1999). Lesbians, men-women, and Two-Spirits: Homosexuality and gender in Native American cultures. In E. Blackwood and S. E. Wieringa (Eds.), *Female desires: same-sex relations and transgender practice across culture* (pp. 91–117). New York: Columbia University Press.

Malinowski, B. (2005 [1929]). *The sexual life of savages in north-western Melanesia.* Whitefish: Kessinger Publishing.

Martin, K., & Vorhies, B. (1975). *Female of the species.* New York: Columbia University Press.

Mead, M. (1962 [1950]). *Male and female. A study of the sexes in a changing world.* Harmondsworth: Penguin Books.

Mitchell, K. (1997). Different diasporas and the hype of hybridity. *Environment and Planning, 15*(5), 533–553.

Murray, S. (1994). On subordinating native American cosmologies to the empire of gender. *Current Anthropology, 35*(1), 59–61.

Ortner, S. (1981). Gender and sexuality in hierarchical societies: The case of Polynesia and some comparative implications. In S. Ortner and H. Whitehead (Eds.), *Sexual meanings: The cultural construction of gender and sexuality* (pp. 359–409). Cambridge: Cambridge University Press.

Padgug, R. (1979). Sexual matters: On conceptualizing sexuality in history. *Radical History Review, 20,* 3–24.

Papastergiadis, N. (1997). Tracing hybridity in theory. Debating cultural hybridity: Multi-cultural identities and the politics of anti-racism. In P. Werbner, & T. Modood (Eds.), *Debating cultural hybridity: Multi-cultural identities and the politics of anti-racism* (pp. 257–281). London: Zed Books.

Pike, K. (1962). *With heart and mind.* Grand Rapids, Michigan: Wm. B. Eerdmans Publishing Co.

Roscoe, W. (1994). How to become a berdache: Toward a unified analysis of gender diversity. In G. Herdt (Ed.), *Third sex third gender. Beyond sexual dimorphism in culture and history* (pp. 329–372). New York: Zone Books.

Roscoe, W. (2000). *Changing ones.* New York: St. Martin's Press.

Rutherford, J. (Ed.). (1990). *Identity: Community, culture, difference.* London: Lawrence and Wishart.

Spivak, G. C. (1988). Can the subaltern speak? In C. Nelson, & L. Grossberg (Eds.), *Marxism and the interpretation of culture* (pp. 271–313). Urbana, Chicago: University of Illinois Press.

Tafoya, T. (1992). Native gay and lesbian issues: The two-spirited. In B. Berzon (Ed.), *In positively gay: New approaches to gay and lesbian life* (pp. 253–259). Berkeley, CA: Celestial Arts.

Towle, E. B, & Morgan, L. (2002). Romancing the transgender native: Rethinking the use of the "third gender" concept. *GLQ: A Journal of Lesbian and Gay Studies, 8*(4), 469–497.

Trexler, R. (2002). Making the American berdache: Choice or constraint? *Journal of Social History, 35*(3), 213–236.

Werbner, P., & Modood, T. (Eds.). (1997). *Debating cultural hybridity: Multi-cultural identities and the politics of anti-racism.* London: Zed Books.

Weston, K. (1993). Lesbian/gay studies in the house of anthropology. *Annual Review of Anthropology, 22*(1), 339–367.

Weston, K. (1998). *Long slow burn: Sexuality and social science.* London: Routledge.

Young, R. (1995). *Colonial desire. Hybridity in culture, theory and race.* London: Routledge.

6 The Fifth Other

In 1968, the psychoanalyst Robert Stoller argued that intersexualized people do not really belong to the human race (1968: 34, see also Chapter 2 in this volume). Sex_gender is constructed as so fundamental for the intelligibility of the human being that in the case of intersex*, gender becomes genus, meaning the human species (Butler, 1993). In this case, the term 'genus' is used in relation to a biological classification, which is ranked below the term species that refers here to 'human being.' In this biological logic, everything that is almost or not anymore human—is degenerate. Degeneration is a term that is charged with cultural and socio-political meaning since the nineteenth century with its emerging colonial, sexological, psychoanalytical, and evolutionary discourses. Anne McClintock describes degeneration as a social figure, rather than a biological concept, which is linked to the idea of contagion and fears concerning "fallibility of white male and imperial potency" (1995: 47).[1] According to Steven Angelides (2001) in the concept of bisexuality—the epistemological sibling of intersexuality—a multiplicity of different notions such as atavism, degeneration, and arrested development were unified. In their unification, they reaffirmed the evolutionary logic of the political differentiation between civilized and primitive evolutionary entities.

The title of this chapter comes from a term coined by anthropologists Andrew and Harriet Lyons in *Irregular Connections* (2004) to denote how anthropologists construct the other as other than white/Anglo/Northern Europe and other than heterosexual. The fifth other is an idea based on Foucault's analysis of the four privileged objects of knowledge. These objects of knowledge are "the hysterical woman, the masturbating child, the Malthusian couple, and the perverse adult" (Foucault, 1978: 105). These four objects are used for the validation and naturalization of the heterosexual, procreating, Christian couple. This chapter, therefore, develops a critique of the location of sexual truth—the heterosexual matrix—in bodies and its accompanying cross-cultural frame of reference, which wants to make sex "the explanation of everything" (78) including the inferiority of non-European societies. All these arguments served the justification of the colonization of non-European cultures by European and US-American nations. Foucault neglected to look explicitly at the construction of a fifth sexual other—the differently sexed savage (Lyons & Lyons, 2004: 101). This analytical lack has led to ignorance towards the interconnection of discourses on race/ethnicity, sexuality, and gender (see also Stoller, 1996). In cross-cultural intersexualization, I see the narrative of a continuum of sexual difference in the body and the degrees of racial or cultural difference merge into the constructions of this fifth other, the differently sexed savage, which serves to consolidate the hetero- and monosexual, white individual.

The roots of cross-cultural intersexualizations are to be found at the intersection of anthropology, sexology, and psychoanalysis in the nineteenth century. In anthropology of the nineteenth century, evolutionary notions helped construct the figure of the so-called sexualized savage that is frequently the object of anthropology's curiosity. Sexology began equally developing as an academic discipline at the end of the nineteenth century—its discourses, terminologies, and metaphors also relied heavily on evolutionary notions—it also has to be seen in the light of the colonial world order which was dominant at the time—just as anthropology. Psychoanalysis a similarly young discipline was occupied with the development of the psyche but also with the development of society—civilization and the orders of 'mature sexuality' in particular.

In the anthropological account I want to focus on, 'the referent of the West' is reiterated through the use of some of the most powerful discourses of the West—psychoanalysis, sexuality, and evolution. The anthropologist Gilbert Herdt is a well-known academic in his field, and a pioneer in introducing the issue of sexuality to his discipline. Herdt, famous for his research in Papua New Guinea, created the concept of ritualized homosexuality (1984) and in the course of his research also became interested in intersexuality (Herdt, 1990, 1994; Herdt & Stoller, 1985).[2] Herdt's own extensive research on sexuality, and on figures such as the sexualized_gendered 'Third' (discussed in Herdt, 1994), occupies a crucial space in anthropological research on sexuality in general, and intersexuality in particular. It is this specific anthropological account of intersexuality as well as its discursive underpinnings, which had major implications for future research into intersexuality and are tied up in specific discourses which are still present in cross-cultural research into sexuality. It is not just that the 'Third'—sex or gender—has to be understood as the addition to the first and second sex that is men and women—even if we do not use Simone de Beauvoir's notion of the 'second sex,' the 'Third' is the addition to the dualism and thereby reaffirms it and its heteronormative framework.

The fifth other: the differently sexed savage emerged through different, historical, discursive preconditions upon which contemporary Western anthropology mostly unwillingly, here represented by the anthropologist Gilbert Herdt (see Chapters 4 and 5 in this volume) based their research. I identify the narratives and modes of representation in ethnographic work on gender, sex, and sexuality that contribute and underpin current rationales for cross-cultural intersexualization. As such, I interrogate the language that is applied to formulate these claims, and pose questions about the translatability of the cultural and symbolic systems of 'other' cultures. I focus on the issue of intersex* in questioning how and when Herdt has decided to speak about intersex* in 'the Others.' By interrogating the interconnected workings of anthropology, sexology, psychoanalysis, I especially engage with the use of psychoanalytical concepts in cross-cultural research, particularly in my readings of the notion of development and degeneration. I will hereby focus on the concept of the polymorphous perverse in Herdt's work, which he used to describe the sexuality of the Papua New Guinean people he researched.

Cross-cultural intersexualization as analyzed in Chapters 4 and 5, and the Excursus contribute to the construction of the fifth other. Contemporary anthropologists such as Gilbert Herdt, just like some nineteenth-century sexologists, for example, refer to hermaphroditism/intersexuality as a third sex and/or third gender. Drawing from psychoanalytic and sexological discourses, Herdt's studies in Papua New Guinea produced an anthropological narrative of the intersexualized other as variably 'savage,' 'naïve,' and 'primitive.' This chapter samples the history of the discourses in sexology, psychoanalysis, and anthropology in order to show the underpinnings of cross-cultural intersexualization produced intersections of racialization and sexualization.

Anthropology and its Referents

Contemporary anthropologists have problematized ethnography as the method of anthropology. James Clifford, for example, argued that ethnographic representations are always 'partial truths;' yet, these partial truths are also 'positioned truths' (Abu-Lughod, 1991: 142). Ethnographers (e.g., Clifford & Marcus, 1986; Geertz, 1973) are, according to Elspeth Probyn, still "united in their use of ethnography as a means of constructing a fundamental similarity of the world's cultures which is firmly based in the referent of the West" (1993: 78). Here, this 'referent of the West' is at stake, together with its diverse modes of reinstalling itself as the center and the 'other' as 'lacking' (relative to the West). The concept of 'the Other.' which is also at stake here, is derived from the works of Luce Irigaray (e.g., Irigaray 1985; Irigary & Guynn, 1995) and Stuart Hall (e.g., 1997). Irigaray writes, from a feminist perspective, on the "fundamental model of the human" which is "one, singular, solitary, historically masculine, the paradigmatic Western adult male, rational, capable. [...] The model of the subject thus remained singular and the 'others' represented less ideal examples hierarchized with respect to the singular subject" (Irigaray & Guynn, 1995: 7). Stuart Hall shifts this perspective slightly, describing a form of racialized knowledge of 'the Other' with reference to Edward Said (1978) and Franz Fanon (1986) who, using the concept of 'Orientalism,' have shown how the hegemonic construction of the white subject is always based on the construction of another non-white model: 'the Other.' The fifth other as it emerges in the combination of both sexualizing and racializing processes reflects these mechanisms.

Psychoanalytical and anthropological discourses, and their intrinsic evolutionary framework in nineteenth-century sexology construed the homosexual and/or the hermaphrodite as an abnormal 'invert' (with regard to sexual dimorphism as the achievement of civilization). The construction of this sexually abnormal invert is analogous to racialized discourses that position non-Western cultures as primitive and less developed. The heritage of these discourses in anthropology, and how they merge into cross-cultural intersexualization still persists at the end of the twentieth century. I see the process of intersexualization as the quest for a scientifically verifiable distinction between men and women. Intersexualization is, therefore, at the core of the process of the construction of a dichotomously sexed_gendered society. Historically, the category of intersexuality has been, and continues to be, formulated as a distinction between male and female and masculinity and femininity (e.g, Fausto-Sterling, 2000; Holmes, 2000; Kessler, 1998), and as a distinction between homo- and heterosexuality (e.g., Adkins, 1999; Butler, 1990; Foucault, 1980). Yet, in the case of cross-cultural intersexualization, we find another distinction mediated through gender and sexuality—the distinction between the civilized and the primitive. In cross-cultural intersexualization, the 'immaturity' of the intersexualized body—the other to the two sexes_genders—stands for the 'immaturity' of the 'other' culture. The terms applied in this twofold othering process vary, yet the notion of development and maturity—concerning the psyche, the body, and a specific culture as a whole is ingrained in the discourses that produce cross-cultural intersexualization.

The Third and its Analogies

Anthropologists Harriet Lyons and Andrew Lyons describe Gilbert Herdt's *Third Sex, Third Gender: Beyond Sexual Dimorphism in Culture and History* (1994) as an "extremely influential volume" (2004: 297), in which Herdt brings together various accounts of so-called 'Thirds' through time and space. This volume is unquestionably an "excellent stimulus to further work along this path" (Conway-Long, 1995: 711). Kath Weston and Morgan Holmes both praise the collection for avoiding the traps in Western notions of what sex this so-called Third 'really' is (Holmes, 2004: 4; Weston, 1993: 349). The title of Herdt's edited book, however, can be read as symptomatic of recent developments in ethnological/ethnographical cross-cultural approaches to sex-gender-sexuality-systems. The volume contains a variety of accounts of sexualized and gendered identities, in different historical periods and across different geographical sites: analysis ranges from the Byzantine period, to sexology at end of the twentieth century, and from the Balkans to India and Polynesia. In the preface to this collection, Herdt states that "the hermaphrodite, for instance, may become a symbol of boundary blurring: of the anomalous, the unclean, the tainted, the morally inept or corrupt, indeed, the 'monsters' of the cultural imagination of modern Americans" (1994: 17). Yet, in "Mistaken Sex," Herdt's own chapter in the collection, he works against this characterization of the hermaphrodite as a 'symbol of boundary blurring,' repeating a common maneuver by explaining cultural and individual expressions through the framework of psychosexual development. The concepts underpinning this framework are derived from sexological, psychoanalytical, and evolutionary discourses originating in the nineteenth century. The consequence of which are to be seen in the implicit notions emerging in cross-cultural intersexualization. Cross-cultural intersexualization therefore, entails the biological essentializing of tri-morphic sexual difference. This complex frame of reference presents another dichotomous component: the construction of ethnicized and racialized psycho-sexual difference. In his article on "Mistaken Sex" Herdt engages in an othering process, both at the level of the subject and the culture.

There is no doubt that the initial motives of ethnologists to interrogate so-called third sexes and third genders incorporated the desire to depict 'other' cultures adequately, and that they were searching for accurate terms to describe their findings. However, the question remains whether this is possible at all. Morgan Holmes, in her article "Locating Third Sexes," has noted that "caution is necessary when culturally specific symbolic orders are employed to prove a(ny) point about Western sex_gender systems; the notion of learning from 'other' cultures raises serious problems" (2004: 5). Holmes states that ethnological research into third sexes_genders is likely to fall into the trap of idealizing cultures which are thought of as representing a version of a symbolic order, to be seen as superior to the limited Western dichotomous conceptualization of sex and gender (2). Furthermore, Holmes criticizes Herdt's collection for "lumping all the erotic and symbolic elements of these cultures together under one rubric of 'third sex and gender' categories" (5). She sees this as a sign that many anthropologists still think "along a dimorphic axis, permitting the occasional disruption to be entertained," but fail to consider that the so-called 'third' might be a 'first' or even "one of any of a multiplicity of possible sex categories" (5). The two dimensions, which Holmes criticizes in Herdt's accounts are, first, the hierarchical connotation the Third takes on in relation to the first and second sex, and second, the limitation of the multiplicity of categories through the construction of the Third. To this, I wish to add a third dimension: the implicit construction of the 'other' culture as sexually childlike and developmentally uncivilized, thereby producing a fifth other—that is the third sex_gender. In Western discourse, this third sex_gender, i.e., intersexuality, is constructed as the result of 'arrested development,' and refers to an 'unfinished' embodiment. Yet, anthropological research into intersexuality situates it as cross-cultural, and combines two othering processes—sexual othering and racial/ethnic othering.

The distinction between sex and gender does not solely rest on the binary between man/male and woman/female, but rather, as Sally Markowitz writes, on "a scale of racially coded degrees," which causes sex_gender difference to culminate "in the manly European man and the feminine European woman" (2001: 391). The history of the construction of these racially-coded degrees in the coordinate system of the 'manly European man and the feminine European woman' has already been interrogated by a number of feminist researchers (Markowitz, 2001; Stepan, 1986, 1993; Traub, 1999; Young, 1995). These constructed and coded degrees rely on analogies and interacting metaphors that only work when they are congruent with cultural expectations. One could say that these analogies only work when they suggest new hypotheses; new systems of implications; and therefore new observations (see Stepan, 1986). Stepan elaborates on this process:

> Because a metaphor or analogy does not directly present a pre-existing nature but instead helps construct that nature, the metaphor generates data that conform to it, and accommodates data that are in apparent contradiction to it, so that nature is seen via the metaphor and the metaphor becomes part of the logic of science itself.
>
> (274)

The similarity evoked in these analogies is not something that can be discovered, but rather, is something that has to be established. Scientific texts, as Linda Birke puts it, are like any other text: they draw upon "narratives [that] are culturally available; powerful metaphors and gendered fables" are to be expected (Birke, 1999: 10).[3] In the case of cross-cultural intersexualization, the analogies which are to be found are deeply rooted in colonial histories and the discourses which surround them. As I have demonstrated in Chapter 1 in this volume, John Money has already used the notion of arrested development when explaining intersex as a premature state. Thereby implying that the normal—and completed state in human development is to be either clearly male or female. Anything that does not conform with these standards is declared to be immature and unfinished. Money used evolutionary and biological metaphors and frames of reference. Herdt on the other hand uses psychoanalytical metaphors in order to support his argument on the polymorphous perverse nature of the Sambia in Papua New Guinea.

The Polymorphous Perverse

Herdt (1994) introduces such a 'powerful metaphor' in his description of the Sambian culture in his chapter in *Third Sex, Third Gender*. In his attempt to describe the events in which the *kwolu-aatmwol* (see Chapter 4 in this volume) could emerge as, what Herdt calls, a third sex and/or gender, he searches for the preconditions that could make such a cultural position possible. Herdt is curious about the circumstances under which the *kwolu-aatmwol* achieves his_her meaning in the Sambian culture; he asks how the Sambian culture could make 'androgyny' a significant motif in cultural representation. Herdt answers his own question, with the help of the Freudian polymorphous perverse. He states that:

> polymorphous cultures such as those of the Sambia of Papua New Guinea, by contrast, define persons as more fluid and as relatively male or female, according to social and development characteristics such as lifespan stage, socioeconomic status, and body ritual.
>
> (Herdt, 1994: 425)

Herdt applies the psychoanalytic term polymorphous perverse to this otherness he detects in the social construction of the Sambian culture, and states that "their permissiveness might be characterized, to use Freud's (…) famous phrase, as 'polymorphous perverse' " (Herdt, 1990: 435). The concept of the poly-morphous perverse in Freudian terms describes a psychological state of being, an early stage of development, before the infant enters into culture or the sym-bolic order. The resolution of the Oedipus complex guarantees that the child becomes a sexualized_gendered being and therefore intelligible. In *Civilization and its Discontents* (Freud, (1961 [1927])), Freud describes the painful process in which civilization chooses certain body parts and makes them represent so-called sexual difference, as well as the use of those parts to justify the only per-mitted sort of love and bodily unions and pleasure which is "heterosexual genital love." This, for Freud, most normal sort of sexuality is an achievement of an albeit never quite complete evolution. Interestingly, in this account, women become more prone to sexual sensitivity because they have two "sex_gender zones" that can give them pleasure. Man supposedly, however, has only one "sex_gender organ" which makes him more unisexed/-gendered.[4] Therefore, women have a greater tendency towards the polymorphous perverse. Freud talks about the child and "das unkultivierte Durchschnittsweib," which is translated as the "average uncultivated woman" which implies that class and race play a big role here in suppressing the polymorphous perverse (Freud, (1961 [1927]: 97). Every woman has the potential to become a prostitute and therefore poly-morphous perverse if not properly cultured into patriarchal, misogynist, heter-onormative society. In the polymorphous perverse there is undifferentiated possibilities of pleasures (and embodiment), which the subject learns to contain and control according to societal, and I want to add, political censure, rules, and requirements. In this sense, everything that remains polymorphous perverse is aberrant and deviant in terms of sexuality, race, and class—be it women or chil-dren (see also Chapter 1 in this volume).

Therefore, in the cross-cultural context of the anthropological realm, the poly-morphous perverse takes on a problematic position. First of all, Herdt ignores the fact that the Euro-American world also includes and recognizes intersexual-ized people, who are regarded as differently sexed. What is more, some people claim an intersexualized or transgendered identity, and live in subcultures where they are perceived as such: in other words, there are spaces in the West which are also 'permissive' to sexual variance. However, in contrasting these two cul-tures that are supposedly so different, Herdt homogenizes not just the culture of the Sambia, but also his own Western culture. Second, while Herdt deploys the Freudian metaphor, he fails to explain why the Melanesian society is supposedly polymorphous perverse—is it due to their bodies, their desire, their gender system, or their otherness? In cross-cultural intersexualization, and especially in the case of Herdt's work, all these combine in the process of othering and produce a sexually and racial 'Other.' Instead, these othering processes are dependent on each other and join forces in the case of cross-cultural intersexualization.

By referring to the term polymorphous perverse, which clearly denotes a state of child development (regarded as prior to 'mature civilization'), Herdt describes the socio-cultural system of the Sambia *in relation* to civilization. He constructs their culture (as a whole) in a developmental psychoanalytical framework, and understands it through a cross-cultural analysis. Herdt, I argue, falls into the trap of one of the most common racializing/ethnicizing analogies; that of positioning of another culture at a stage that is less developed, more childlike and primitive, in relation to the civilized, sophisticated, and developed Western/Christian/ Global North civilization. This evokes common themes of 'the West' as superior, and of 'the Other' as defined by lack. Following Neville Hoad, I would argue that the processes of cross-cultural (homo-)sexualization are not understandable without looking at the imperial and neo-imperial contexts of such theoretical productions (Hoad, 2000). The processes of cross-cultural intersexualization are comparable to Hoad's emphasis on the emergence of the homosexual through a "hierarchical staging of human difference under the historical period of imperialism and globalization and the attendant logics of evolution and development respectively" (133). This notion combines theories of the body with theories of the mind to jointly constitute this "one signifier" by "progress through its various others, which are then posited as vestigial, arrested, anachronistic or degenerate" (134). Whenever, the differently sexed savage is constructed in cross-cultural intersexualization, two interdependent processes are at work: racialization and sexualization. Both processes pathologizes that which is not 'fully' developed—the one as civilized and the other as heterosexual.

Herdt's representation of this particular culture as permissive, and therefore actually progressive, with regards to sexual variation, invokes Freudian psychoanalysis as a universal discourse. Even though the Sambian culture Herdt discusses does not organize itself in the same manner as Freudian nineteenth-century bourgeois Vienna, and does not share this history, he uses the concept of the polymorphous perverse to describe the psychological processes, and the organization of sexuality, of the Sambian culture. However, the connotation of this concept works against the argument Herdt actually wants to make, because the polymorphous perverse implies lack (of sexual differentiation) with regards to civilization *and* sexuality, and therefore forecloses the perception of the culture as progressive.

Psychoanalysis as a (Neo-)colonial-Evolutionary Discourse

Sigmund Freud's theories, developed in *Totem and Taboo* (1953 [1913]), relied, as postcolonial theorist Kalpana Seshadri-Crooks notes, "on the parallels between primitives and neurotics" (1994: 190). Freud conceptualized so-called primitive cultures' minds as fundamentally different to the thinking of the logocentric West. To him, the 'primitive mind' does not differentiate the mystical from reality; rather, it uses 'mystical participation' to interpret and manipulate the world. As such, Freud invests in imperialist and colonial discourses, which were present in his time; the mind of the 'savage' had a specific function in anthropological discourses at the turn of the twentieth century. For example, Lucien Lévy-Bruhl (1857–1939), a philosopher and so-called armchair anthropologist, published various texts on the 'primitive mind' and the "essential difference between the primitive mentality and ours" (Lévy-Bruhl, 1975: 4). Seshadri-Crooks notes that "the difference between the savage and the civilized man is expressed on a diachronic axis, as a temporal difference in 'our past' and is not subject to an interchangeability of the actors" (1994: 195). The notion of development towards is here even extended to the development language implying that communication in Western cultures used to be just as it is now in so-called uncivilized cultures.

In *Dark Continents* (2003), postcolonial theorist Ranjana Khanna goes even further, examining the embeddedness of psychoanalysis in geo-political and historical coordinates and understands psychoanalysis as a colonial discipline. Her arguments conceptualize psychoanalysis as an ethnography of nation-statehood, and examine the impossibility of adequately understanding psychoanalysis "without considering how it was constituted as a colonial discipline through the economic, political, cultural, and epistemic strife in the transition from earth into world" (9). This allows her "to see how nation-statehood for the former colonies of Europe encrypts the violence of European nations in its colonial manifestations" (6). Khanna argues for a provincializing, politicizing and historicizing of psychoanalysis, to counter the intrinsic universalizing motions derived from its geo-political and historical origins. This is needed not only for anthropological psychoanalysis, but also for a re-examination of some basic texts by Freud, such as *Totem and Taboo* (1953 [1913]) and *Civilization and its Discontents* (1961 [1927]), which rely heavily on the distinction between the civilized world, the Western capitalist nation states, and the so-called 'savage' societies. For example, in *Totem and Taboo* (1953 [1913]), Freud borrows from a theory of homology that assumes that ontogeny recapitulates phylogeny: this means that the individual repeats the stages of the development, or the evolutionary stages of the species. According to Khanna, psychoanalysis cannot be considered without inclusion of the "evolutionary logic that informs Freud's sense of the growths of repression in civilization" (2003: 11). This evolutionary logic unfolds especially in Freud's theories on psychosexual development.

Not only is the notion of repression enabling civilization problematic, but so too is the evolutionary logic found in Freud's theories on psychosexual development. In 1939, Freud admitted that "I must, however, in all modesty confess that (...) I cannot do without this fact in biological evolution" (100). Indeed, Freud heavily relied for example on Lamarckian theories to support his claims about so-called psychosexual development. He also depended on Charles Darwin's theories about the arrangement of early human societies, thus locating the beginnings of the Oedipus complex in the origins of human society.[5] Examining the Freudian concept of psychosexual development, the resolution of the Oedipus complex appears to be the stage in which the child, and culture, leaves the child-like and generic form of an uncivilized being behind, and emerges into a 'mature' organization between self and others. Freud enveloped the psychic and social in an evolutionary rhetoric. By referring to Johannes Fabian's classic *Time and the Other: How Anthropology Makes its Object* (1983), Neville Hoad states that the social evolutionists "discarded Time altogether." Moreover, "the temporal discourse of anthropology as it was formed decisively under the paradigm of evolutionism rested on a conception of Time that was not only secularized and naturalized but also thoroughly spatialized" (Hoad, 2000: 135). Here, the paradox regarding the uses of time demonstrates the construction of development as highly geo-political. Underlying this construction are evolutionary theories that date back to the nineteenth century. As the child of colonialism, anthropology is immersed in these theories, and as such, has significantly contributed to a hierarchical ordering of the world. The notion of development towards is crucial in this process. By temporalizing space, contemporaneous non-European cultures become understood as the representatives of Europe's past. Through this model, the possibility of understanding cultural difference is precluded, because it insistently implies that the 'civilized' Western culture has already been the 'primitive,' the non-Western (Hoad, 2000: 142).

Gender Studies scholar Steven Angelides states that the publication of Charles Darwin's *On the Origin of Species* in 1859 "effectively canonized evolutionary thinking, leaving few spheres of Western thought untouched" (2001: 29).[6] The origins of psychoanalysis and of anthropology, especially anthropology concerned with sex, gender, and sexuality, indeed demonstrate traits of evolutionism. The simultaneous development of the two basic, but rival theories of diffusionism and evolutionism in the nineteenth century created debates about the differing underlying theoretical frameworks. Cultural anthropologist Jack Stauder describes the anthropological tradition around the turn of the twentieth century as being "dominated by controversies between diffusionists and evolutionists who held in common, however, an historical and often speculative approach that was primarily concerned with reconstructing the past of mankind" (1993: 409). Whereas the diffusionists were interested in tracing wildly dissimilar societies back to commonly shared cultural origins and connections, the evolutionists relied on a theory of linear and separate development of societies. This notion of a linear but separate development was based on Darwinian narratives of evolution. Applied to the development of human societies, this produced narratives about the evolution of humankind, ranging from 'savagery' and 'barbarism' to 'civilization,' on an evolutionary continuum that can be seen across different cultures. The conclusion drawn in the imperialist era at the end of the nineteenth century was that 'advanced' societies have the responsibility of civilizing 'primitive' societies. I argue that the anthropological power/knowledge complex crystallizes in its most material form when linking the 'past of mankind' with the evolutionist explanatory framework concerning sexuality.[7]

Ann McClintock elaborately describes this 'vital analogy' of (arrested) development and non-European cultures in evolutionary theory. According to her, we can assume that if the "white child was an atavistic throwback to a more primitive adult ancestor," this child "could be scientifically compared with other living races and groups to rank their level of evolutionary inferiority" (1995: 50). The adults of inferior groups ('savage cultures,' 'non-sophisticated' societies, etc.) must be like the children of superior groups (industrialized societies); in this analogy, the child represents a primitive adult ancestor who is thought of as being in the same stage of mental development as the adult of the so-called savage society (50). Stephen Gould relates this to racialization, stating that "if adult blacks and women are like white male children, then they are living representatives of an ancestral stage in the evolution of white males" (1981: 115). He concludes that "an anatomical theory of ranking races—based on entire bodies had been found" (115). Gayatri Spivak, questioning the entire foundation of scientific knowledge production, states that "in fact, if the analogy between primitive peoples and children were not scientific, the fundament of the science would be blown away" (1993: 20). I suggest that Herdt's use of the metaphor of the polymorphous perverse is the foundation for his claim about the permissiveness of the other towards sexual variation intelligible to his Western audience. However, as I argue, the use of this metaphor is further consolidated when cross-cultural intersexualization is at work in the same maneuver.

The influence of Darwinist ideas on categorizations of sex, gender, and sexuality has already been widely discussed (e.g., Somerville, 1994; Hoad, 2000). Siobhan Somerville notes that one of the basic hypotheses within Darwinian thinking was that organisms evolve through a process of natural selection and, therefore, also show "greater signs of differentiation between the (two) sexes" (1994: 255). The notion of sexual dimorphism as the pride of evolution, and therefore civilization, is central to intersexualization. In this evolutionary narrative, racialization features even before Darwinism gained influence. Imperialism and colonialism existed before the end of the nineteenth century, and already needed justification through the construction of the 'inferior other' who can be exploited/extinguished without further explanation; the inferiority of the other made any explanations redundant. Moreover, the trafficking between cultures and continents endangered the purity of the civilized white 'race;' miscegenation was a trope that began to cause anxiety. The notion of 'mixed-breeds,' so-called hybrid products of a marriage between a 'white' and a 'non-white' person, is deeply intertwined with nineteenth century eugenicist and scientific-racist discourse (Young, 1995; Mitchell, 1997; Werbner & Modood, 1997).

Therefore, in the nineteenth century, sex and race increasingly came to define social value. Anatomists from this period studied sex and race, and according to Londa Schiebinger, positioned the European white male as the "standard of excellence" (1989: 212). The analogy between sex and race, as Schiebinger suggests, has drawn on a variety of reference points, at different points in time, since the eighteenth century. The early framework of the production of racial differences was rooted in anatomy, which molded differences into muscles, nerves, and veins. Like sex, race, Schiebinger concludes, came to penetrate the "entire life of the organism" (211). According to Markowitz, with the beginnings of sexology, the focus shifted to measurements of the pelvis (2001), which was thought to be equally important for understanding the physical and moral development of the 'races.' Schiebinger states that "with pelvis size, sexual (although not racial) hierarchy was reversed. Here the European female represented the fully developed human type, outranking the European male" (1989: 212).

However, this did not mean that European women became the superior 'species;' they just became, in a eugenic framework, the best choice for the white man for procreation, reaffirming "blonde heterosexuality" (Markowitz, 2001: 404).

Sexologists focused on the pelvis as the body part that revealed the physical and moral development of the 'races.' As many historians have documented, nineteenth-century European thought was preoccupied with women's reproductive ability, their (uncontrollable or frigid) sexuality, and their (hysterical) psychology (Gross & Averill, 1983: 81). Darwin provided the framework to justify the political, legal, and cultural asymmetry between men and women (particularly necessary in light of feminist movements that threatened to disrupt the traditional patriarchal organization of society). In the Darwinian model, sexual behavior centered on reproduction. Social Darwinism, fostered by the publications of Herbert Spencer and others argued that "it is in the struggle for existence and (especially) for the possession of women that men acquire their vigor and courage" (Angelides, 2001: 30). Within sexology, heterosexuality was the teleological, necessary, and highest form of sexual evolution. Sexologist Havelock Ellis suggested in 1911 that "since the beginnings of industrialization, more marked sexual differences in physical development seem (we cannot speak definitely) to have developed than are usually to be found in savage societies" (13). Similarly, Iwan Bloch states in 1907 in *The Sexual Life of Our Time* that with civilizational progress the contrast between the sexes becomes "continually sharper and more individualized" (58). According to Bloch, "primitive conditions" found in other cultures and "in the present day among agricultural laborers and the proletariat" in Europe, create sex(ual) difference as "less sharp and to some extent even obliterated" (58) The achievement of sex(ual) difference as a sharp contrast between the sexes is the white middle-class lady and gentleman of the 'civilized' world as the pinnacle of achievement. This contrast, however, needs to be literally mediated by a figure which lies 'in-between' the two parameters of sexual difference and racialized/ethnicized—here 'class' is also at stake—to make these two continua intelligible.

In 1896, the homosexual author, Edward Carpenter discusses the principle of "the third" in *The Intermediate Sex* (1921). Carpenter resisted scientific associations between homosexuality and degeneration, *The Intermediate Sex* attempts to free the 'intermediate sex' (that is, homosexuality) from pathology and abnormality. Carpenter suggested that 'intermediary types' exist on a continuum 'in-between' the poles of exclusively heterosexual male and female. Using rhetoric from the racial sciences, he argued that 'shades' of sexes and sexual 'half-breeds' assigned homosexuality a place in the natural order. The analogy between the 'sexual invert' and the 'mixed racial body' was employed in contradictory ways. On the one hand it could assign the homosexual a legitimate place within the natural order; on the other hand, it could demonstrate degeneration (Somerville, 2000: 33). Carpenter stated that "anatomically and mentally we find all shades existing from the pure genus man to the pure genus woman" (1921: 133). This became the central feature of the continuum of the natural order in which the 'pure' bodies of white heterosexual men and women were positioned at the far end of civilization by reference to the 'natural' developmental stages of 'in-between.' Anne McClintock explains the social power of the image of degeneration by referring to the description of social classes or groups as "races," "foreign groups," or "nonindigenous bodies," which "could thus be cordoned off as biological and 'contagious,' rather than as social groups" (1995: 48). McClintock concludes that the usefulness of the quasi-biological metaphors of "type," "species," "genus," and "race" was that they gave "full expression to anxieties about class and gender insurgence without betraying the social and political nature of these distinctions.

In their highly influential and widely read 1889 publication, *The Evolution of Sex,* Patrick Geddes and Arthur Thomson state that "hermaphroditism is primitive; the unisexual state is a subsequent differentiation" (1889: 80). The notion of natural selection made it possible to view hermaphroditic/intersexualized bodies as anomalous evolutionary 'throwbacks.'[8] Referring to this history, Ulrike Klöppel states that hermaphrodites were therefore regarded as "atavistic monstrosities" (2002: 161). Foucault (2003) demonstrates that during the nineteenth century, the hermaphrodite was placed in the category of a 'monster' that disrupted the whole intelligible order and rationalizing apparatus. The assignment of meaning to certain identities, or the construction of these identities in the first place, cannot be detached from the subsequent assignment of a place in the order of beings for these newly created identities. This order is hierarchically configured, and derives its parameters from the discourses that explain what human nature or the human species is supposed to be—who can be included and who cannot. The othering processes we find in this literature are already two-fold; they need each other to be made intelligible.

In these accounts we find the polymorphous perverse hidden, yet emerging through underlying concepts such as development (cultural or sexual). In several instances, the psychological, sexological, and the evolutionary discourses interlink and produce the Freudian notion of the polymorphous perverse which will be transformed into the sexual *and* racial 'other' in Herdt's work nearly a century later. However, the etymology of gender and genus, as well as the interconnected epistemologies of bisexuality and intersexuality demonstrate even more thoroughly the conceptual origins of cross-cultural intersexualization.

Genus and Gender

Thus, the hermaphrodite came to be seen as atavistic, and as unfinished in its development. The term, degeneration, as related to evolutionism, also entered the debate. Havelock Ellis noted that conflating the homosexual and the hermaphrodite in one term was common for sexologists, stating that "strictly speaking, the invert is degenerate" (Dreger, 1998: 138). Alice Dreger claims that Ellis disliked the term, and made it clear that he only used it in the "most scientific sense," which meant that the hermaphrodite "has fallen away from the genus" (Dreger, 1998: 138). 'Genus,' a Latin term, means 'race' or 'kind.' 'Degeneration' derives from the Latin word 'generare,' which means to procreate or breed, but also to generate, to foster, and to produce. It implies deterioration from the norm, in terms of being the type of human being considered to be the norm, and being a (re)productive member of society. Therefore, to degenerate, or to be degenerate, means to not belong to the human race, but also to not be generative or productive. Degeneration is inexorably linked to development and maturity.

The notion of the hermaphrodite as having 'fallen away from the genus' means that intersexualization functions through an exclusion from the norm of the human species, and from its subdivision 'genus = gender.' In the sexological discourse of Ellis's time, this implies that the 'invert' (here, standing for the homosexual and the hermaphrodite) could not, or rather should not, reproduce; not only because of the invert's negation of reproduction made consistent through the heterosexual matrix, but also because degeneration is inheritable and is also intrinsically linked to eugenics (Barnett, 2006). Sander Gilman argues that sexuality is the most salient marker of otherness, organically representing racial difference (1985, 1993). I argue that intersexuality serves to organically represent racial/ethnic difference. The interconnection of the notions of degeneration and the 'human race' is made comprehensible in terms of a continuum of sexual dimorphism and racial/cultural difference. This continuum combines the discourses of racialization/ethnicization in intersexualization. Ellis, Bloch, and Carpenter were not the first ones to draw on this interconnection to make their sexological theories intelligible to their contemporaries. The categories of race, class and sex_gender, as well as sexuality, are not structurally equivalent; however, through analogy and metaphor, they are co-constructs in scientific discourses. Their historical heritage feeds into current conceptualizations of cultural and ethnic difference, and informs interpretations and explanations of the body, desire, and difference.[9] The interconnection and/or analogies which Ellis and his contemporaries built on are based on a tradition that dates back to the Enlightenment and the beginnings of imperialism (and, therefore, also anthropology as an academic discipline) (Schiebinger, 1993; Stepan, 1996). Not only did the material body have to bear theories of inferiority and degeneracy, but the categories of 'morality' and 'social worth' informed and underpinned these theories, mainly through craniology—the study of the scull. This branch of science is only intelligible if social categories are added to theories about the differences between racialized and sexualized bodies and identities. Hoad has called this process the "reinscription of biological evolutionism into the sphere of the psychic" (2000: 141). In cross-cultural intersexualization this twofold process is mediated by the notion of development.

Infantile Sexuality/(Neo-)colonial Sexualization

A powerful association between sexual development and 'maturity' emerged from the early theories on psychosexual development and sexuality (and later, on gender identity); Freud emphatically stated in 1905, "every pathological disorder of sexual life is rightly to be regarded as an inhibition in development" (Freud, 1905: 208).[10] Jerome Neu, scholar of Freudian psychoanalysis, notes that "perverse sexuality is, ultimately, infantile sexuality" (1991: 185). In this Freudian sense, infantile sexuality must be understood as a space of non-genital forms of pleasure. Myra Hird states that "perversions are now associated with 'regressed' and/or 'fixated' pleasures rather than mature genital love" (2003: 1075). Further, Neu reads in Freud the collapse of "the individual's experienced concern for genital pleasure together with the biological function of reproduction, so that the development and maturation criterion for perversion reduces to the question of the suitability of a particular activity for reproduction" (187). Neu also refers to the "ideal of maturation," which, according to him, "gives a central role to that function [reproduction] and makes all earlier sexuality necessarily perverse. The infant's multiple sources of sexual pleasure make it polymorphous perverse" (187). Freud moved from conceptualizing homosexuality as a variant of sexual function, to inscribing it as 'arrested sexual development.' Psychoanalysis, even after Freud, draws heavily on most of the evolutionary vocabulary. The notion of 'normal sexuality' became tightly bound to notions of adulthood and 'healthy and mature' development, also with regard to cultural differences. This tying of the concept of sexuality to the notion of development relies on analogies and metaphors, which appear in constructions of cultural (or racial) difference, as well as in constructions of sex(ualiz)ed_gendered difference. Hoad states that "the difference between the perverse and the normal can only be understood in terms of development" (2000: 145). Concerning intersexualization, this statement about homosexuality has to be extended to physical abnormality. I argue that, in cross-cultural intersexualization, the discourses of psychological and physiological abnormality merge when the terms of development are concerned. Moreover, abnormality is construed in evolutionary terms, with cross-cultural reference, in order to create a notion of the normal, mature, and civilized white, Western, binary, hetero-relational matrix comprised of two distinct sexes_genders. The distinctions which are mediated in cross-cultural intersexualization are racial *and* sex(ualiz)ed_gendered boundaries.

Bisexuality and Hermaphroditism/Intersexuality

Whereas Steven Angelides' work demonstrates the place that the concept of bisexuality took in the theories described above, my focus rests on intersexuality. Both categories have, at times, been interchangeably applied or separated. Bisexuality, as Angelides historicizes it, can be regarded as "not unlike the evolutionist's 'missing bisexual link,'" which, just as the hermaphrodite, "served as the dialectical link between the two forces structuring Freud's work: the biological and the psychological" (2001: 53). Angelides states that biological or innate bisexuality, which is hermaphroditism in Freud's understanding, "was Freud's link to the natural sciences" and "epistemologically bisexuality was figured not only as the 'other' to sexual ontology itself, but as the liminal figure through which, and against which, racial, gender, and sexual identities were invented as distinctly separate species of humankind" (2001: 24). Angelides traces this back to the theoretical developments and sums it up as follows:

The universal starting point for all human development, and thus human differentiation, was embryological bisexuality. As children, men passed through physical and psychical stages of bisexuality until maturity, until *(hu)manhood*. Women and blacks, on the other hand, remained children, undeveloped men; or in Irigaray's terms, *sexes which were not ones*. This meant that each of them was therefore a (hu)man that was not one. For it was in the evolutionary process of becoming (hu)man that one was to transcend the physical and psychical animal ancestry of primordial bisexuality. In the Darwinian chain of being, this was an upward movement out of the domain of nature and into that of culture; an evolutionary progression from sexual ambiguity to sexual distinction.

(33 [emphasis in original])

Thus, biological bisexuality—that is, in this account, intersexuality—held a specific place in the ordering of human nature, not just with regard to sex, but also to race/ethnicity. From this time onwards, the notions of 'maturity,' 'arrested development,' 'development,' and the definition of human nature were intrinsically connected. Innate bisexuality was the pivotal epistemic tool, instrumentalized to keep the crisis of white masculinity of the late nineteenth century at bay. Sexologists worked with terminologies and metaphors used by anthropologists, and anthropologists founded a discursive culture based on sexological terminologies. In the processes of cross-cultural intersexualization, these different strands merge and produce a twofold othering process.

Conclusion

Anthropological research and discourse is colored by evolutionary discourses and notions which date back to the nineteenth century, and have provided the foundations for the work of anthropologists. A similar evolutionary discourse is found in theories about intersexuality, in terms of biological and psychological theories. The interdisciplinary agendas of medical and psychological anthropology, particularly when it comes to sex-gender-sexuality-systems, are saturated with new concepts and categories that are invented to apprehend 'the other.' The set of ideas discussed here shaped sexology at the turn of the twentieth century, and continued to inform research into sex, gender, and sexuality as it developed throughout the twentieth century. At the end of the nineteenth century, the acceptance of Darwinism was total; human beings were conceptually connected with the smallest entity[11] and the idea of evolution with Man on top was established—that is to say, anthropocentrism in its most explicit form. Every being was considered to have a place in the evolutionary process of creation: progress was perceived in those species that exhibited the greatest degree of sexual difference, and where heterosexuality was organized around procreation. The notion of development is deeply ingrained in Western research into non-Western societies, where one's own culture and gender regime is set as the highest possible form of development and civilization. The model of development is intrinsically interwoven into the very history of psychoanalysis/psychology, sexology, and anthropology, and therefore in cross-cultural intersexualization.

Lyons and Lyons identify two motivations for anthropological accounts of homosexuality: one, to make available information that has previously been distorted, and two, addressing contemporary gay political issues. However, they state these motivations are "by no means exclusive but are often merged" (2004: 295). They assert that some

> anthropologists are not so much studying the "sexuality of the Other" as implicitly diagnosing "otherness" on the basis of sexuality, even though they have, in many cases, been attracted to their field subjects because of a "sameness" of sexual orientation.
>
> (305)

In this regard, the study of homosexuality is very similar to anthropological cross-cultural intersexualization. Gilbert Herdt's quest to argue for less restrictive and more flexible sex-gender-sexuality-systems produces the 'other' culture as 'other' because of their 'permissiveness.' Applying the metaphor of polymorphous perversity to the representation of the Sambia entails positioning it at the stage of immaturity, of childlikeness. To cross-culturally intersexualize as Herdt does, is to solidify this claim to invoke the notion of the 'incomplete' intersexualized, and construct it as emblematic for the incomplete, or even childlike, primitivism of Sambian culture. In cross-cultural intersexualization, the immaturity of the intersexualized body stands for the immaturity of the culture in which the intersexualized body can exist as such.

In cross-cultural intersexualization the call for acceptance of sexual (biological) variation is made with reference to psychological terminology. By applying the term polymorphous perverse, Herdt evokes the coordinates of arrested development/maturity and savage/civilized. The implicit supposition of non-maturity, in terms of a socially restrictive interpretation of sex-gender-sexuality-systems—and the positioning of the hermaphrodite in this immature organization—produces, as I suggest, 'the Other' as doubly othered. The permissiveness of 'the Other' in the example by Herdt which I interrogated, and an openness to the polymorphous multiplicity of existence, are othered through intersexualization; subsequently, the system that enables this is also othered. Mutual metaphorical affirmations of the two processes of sexual and ethnic othering work towards cross-cultural intersexualization. The pathological characteristic of 'the Others'—their psycho-sexual non-maturity—is reproduced in the singular intersexualized body and in the collectivity of the 'other' culture.

Notes

1. McClintock explains the social power of the image of degeneration by referring to the description of social classes or groups as "races," "foreign groups," or "nonindigenous bodies," which "could thus be cordoned off as biological and 'contagious,' rather than as social groups" (1995: 48). McClintock concludes that the usefulness of

the quasi-biological metaphors of "type," "species," "genus," and "race" was that they gave "full expression to anxieties about class and gender insurgence without betraying the social and political nature of these distinctions. As Condorcet put it, such metaphors made 'nature herself an accomplice in the crime of political inequality'" (48).

2. Herdt was not the first one to interrogate intersex* in other cultures. One of the most famous examples is Robert Edgerton who conducted research in East Africa as early as 1964.

3. Rudi Bleys writes that "historically, the European construction of sexuality coincides with the epoch of imperialism and the two inter-connect" (1995: 106). What lies at the heart of an anthropological configuration of the power/knowledge complex is the "pervasive understanding within anthropology (…) that the human body generates a host of potent metaphorical constructions for ordering the world" (Sharp, 2000: 315). But "the metaphors are inappropriate for translating the concepts of the particular culture: they assimilate alien cultural forms 'too easily' to European [i.e., Western] categories and conceptions" (Street, 1990: 242). The assumption that the other culture under investigation uses the same metaphors or signifiers to designate their peoples' sex_gender is intrinsically colonizing.

4. Freud obviously did not know anything about the prostate, which can be described as a second male sexual organ. The prostate produces part of the semen and is located between the bladder and the rectum. In Western discourse this organ has not been regarded as a sexual erotogenous zone. For thousands of years the prostate is known as a male sexual organ in traditional Chinese medicine or the Tantra.

5. Frantz Fanon, for example, famously disempowered the Oedipus complex as a universally adaptive psychoanalytical structure. Fanon denied the existence of the Oedipus for Martinique, mainly because no black father exists to mirror as The Father and, therefore, no struggle for the mother can take place. He argues that the father is always the White Father, the Colonizing Father—a structural father and not a personal one (Fanon, 1986).

6. Charles Darwin actually returned from his voyages unconvinced that species had emerged through a naturalistic and mechanistic process of evolution. It was not until Darwin read Thomas Malthus's *An Essay on the Principle of Population* (1798) that he found a theoretical construction he could use to frame evolutionary processes in nature. Malthus's political views of the necessity of a "capitalistic defense of middle class accumulation, expansion and domination," as well as the male control of reproduction, found their way into Darwin's theory of evolution (Gross & Averill, 1983: 75).

7. The so-called 'father of anthropology' E. B. Tylor was a crucial figure in establishing evolutionist notions of the development of civilization. He published *Primitive Culture* in 1871 (1958) and *Researches into the Early History of Mankind* in 1865 (1964). Tylor relied heavily on Darwin's theories and often likened 'primitive' cultures to children. To describe the relation between "savage intelligence" and "civilized mental culture," Tylor used tropes from evolutionary theory. He also reasoned that "throughout all the manifestations of the human intellect, facts will be found to fall into their place on the same general lines of evolution" (Tylor cited in Leopold, 1980: 31). The analogy of human evolution and the difference between cultures at the level of the individual and the 'species' became a fashionable rhetorical maneuver in anthropology. Tylor often relied on the standard Enlightenment classifications of societies as 'savage' and 'childlike,' or 'civilized' (Leopold, 1980).

8. Geddes and Thomson discuss Darwin's theory of sexual selection at length in the first chapter of their book (1889: 3–31).

9. McClintock argues that history is not produced around one single privileged social category and that racial and class differences cannot be "understood as sequentially derivative of sexual difference, or vice versa" (1995: 61). To her, the determining categories of imperialism come into being only in their historical relationship to each

other and emerge, in this relationship, in a "dynamic, shifting, and intimate interdependence" (61).

10. The rhetorical gymnastics used to justify surgical intervention in intersex-identified newborns also draws on this notion of development. The parents are not told of the physician's diagnosis, as it is imagined as being too difficult to cope with. This relies on the notion that their child is not fully developed yet, and that physicians must operate in order to secure full sexual differentiation.

11. Over the years and centuries, this smallest entity has become smaller and smaller; now we have reached the level of hormones, chromosomes, and genes. For a critical analysis of genetics see Joan Fujimura 2006, on endocrinology see Nelly Oudshoorn, 2001.

Bibliography

Abu-Lughod, L. (1991). Writing against culture. In R. G. Fox (Ed.), *Recapturing anthropology. Working in the present* (pp. 466–479). Santa Fe, New Mexico: School of American Research Press.

Adkins, R. (1999). Where "sex" is born(e): Intersexed births and the social urgency of heterosexuality. *Journal of the Medical Humanities, 20*(2), 117–135.

Angelides, S. (2001). *A history of bisexuality*. Chicago, London: The University of Chicago Press.

Barnett, R. (2006). Education or degeneration: E. Ray Lankester, H. G. Wells and the outline of history. *Studies in History and Philosophy of Science Part C, 37*(2), 203–229.

Birke, L. (1999). *Feminism and the Biological Body*. Edinburgh: Edinburgh University Press.

Bleys, R. (1995). *The geography of perversion. Male-to-male sexual behaviour outside the west and the ethnographic imagination. 1750–1918*. New York: New York University Press.

Bloch, I. (1907). *The sexual life of our time in its relation to modern civilization*. London: William Heinemann.

Butler, J. (1990). *Gender trouble. Feminism and the subversion of identity*. New York: Routledge.

Butler, J. (1993). *Bodies that matter: On the discursive limits of 'sex.'* New York: Routledge.

Carpenter, E. (1921 [1896]). *The intermediate sex. A study of some transitional types of men and women*. London: George Allen Unwin Ltd.

Clifford, J., & Marcus, G. (1986). *Writing culture. The poetics and politics of ethnography*. Berkeley, London: University of California Press.

Conway-Long, D. (1995). Review: Third sex, third gender: Beyond sexual dimorphism in culture and history, by Gilbert Herdt. *Current Anthropology, 36*(4), 709–711.

Darwin, C. (2003 [1859]). *On the origin of species by means of natural selection, or the preservation of favoured races in the struggle for life*. London: John Murray.

Dreger, A. D. (1998). *Hermaphrodites and the medical invention of sex*. London: Harvard University Press.

Edgerton, R. (1964). Pokot intersexuality: An East African example of the resolution of sexual incongruity. *American Anthropologist, 66*(6), 1288–1299.

Ellis, H. (1911). *Man and woman: A study of human secondary sexual characteristics* (4th ed). New York: Scribner's.

Fabian, J. (1983). *Time and the other: How anthropology makes its object*. New York: Columbia University Press.

Fanon, F. (1986). *Black skin, white masks.* New York: Grove Press.

Fausto-Sterling, A. (2000). *Sexing the body. Gender politics and the construction of sexuality.* New York: Basic Books.

Foucault, M. (1978). *History of sexuality.* London: Penguin Books.

Foucault, M. (Ed.) (1980). *Herculine Barbin: Being the recently discovered memoirs of a nineteenth-century French hermaphrodite.* New York: Pantheon Books.

Foucault, M. (2003). *Abnormal. Lectures at the Collège de France 1974–1975.* New York: Picador.

Freud, S. (1953 [1913]). Totem and taboo. In James Strachey (Ed. and Trans.), *The standard edition of the complete psychological works of Sigmund Freud* (Vol. 13) (pp. ix–163). London: Hogarth.

Freud, S. (1961 [1905]). Three essays on the theory of sexuality. In *The Standard Edition of the Complete Psychological Works of Sigmund Freud* Vol. 7. James Strachey (Ed. and Trans.). London: Hogarth, 125–248.

Freud, S. (1961 [1927]). Civilization and its Discontents. In James Strachey (Ed. and Trans.), *The standard edition of the complete psychological works of Sigmund Freud* (Vol. 21) (pp. 59–246). London: Hogarth.

Freud, S. (1964 [1939]). Moses and monotheism: Three essays. In James Strachey (Ed. and Trans.), *The standard edition of the complete psychological works of Sigmund Freud* (Vol. 23) (pp. 3–140). London: Hogarth.

Fujimura, J. (2006). Sex genes: A critical sociomaterial approach to the politics and molecular genetics of sex determination. *Signs: Journal of Women in Culture and Society, 32*(1), 49–81.

Geddes, P., & Thomson, J. A. (1889). *The evolution of sex.* London: Walter Scott.

Geertz, C. (1973). *The interpretation of cultures.* New York: Basic Books.

Gilman, S. (1985). *Difference and pathology. Stereotypes of sexuality, race and madness.* Ithaca, London: Cornell University Press.

Gilman, S. (1993). *Freud, race, and gender.* Princeton, N. J.: Princeton University Press.

Gould, S. (1981). *The mismeasure of man.* New York: Norton & Co.

Gross, M., & Averill, M. B. (1983). Evolution and patriarchal myths of scarcity and competition. In S. Harding and M. B. Hintikka (Eds.), *Discovering reality. Feminist perspectives on epistemology, metaphysics, methodology, and philosophy of science* (pp. 71–98). Dordrecht, Boston, London: D. Reidel Publishing Company.

Hall, S. (1997). The spectacle of the "Other." In S. Hall (Ed.), *Representation: Cultural representations and the signifying process* (pp. 12–74). London: Sage.

Herdt, G. (1981). *Guardians of the flutes: Idioms of masculinity.* New York: McGraw-Hill.

Herdt, G. (Ed.). (1984). *Ritualized homosexuality in Melanesia.* Berkeley: University of California Press.

Herdt, G. (1990). Mistaken gender: 5-alpha reductase deficiency and biological reductionism in gender identity reconsidered. *American Anthropologist, 92*(2), 433–446.

Herdt, G. (1994). *Third sex third gender. Beyond sexual dimorphism in culture and history.* New York: Zone Books.

Herdt, G. and Davidson, J. (1988). The Sambia "Turnim-Man": Sociocultural and clinical aspects of gender formation in male pseudohermaphrodites with 5-alpha reductase deficiency in Papua New Guinea. In *Archives of Sexual Behavior, 17* (1), 33–56.

Herdt, G., & Stoller, R. (1985). Sakulambei—A hermaphrodite's secret: An example of clinical ethnography. *Psychoanalytic Study of Society, 11*, 117–158.

Herrmann, S. K. (aka S_he). (2003). Performing the gap—Queere Gestalten und geschlechtliche Aneignung. *Arranca! 28*, 22–26.

Hird, M. (2003). Considerations for a psychoanalytic theory of gender identity and sexual desire: The case of intersex. *Signs: Journal of Women in Culture and Society, 28*(4), 1067–1092.

Hoad, N. (2000). Arrested development or the queerness of savages: Resisting evolutionary narratives of difference. *Postcolonial Studies, 3*(2), 133–158.

Holmes, M. (2000). Queer cut bodies. In J. A. Boone, M. Dupuis, M. Meeker, K. Quimby, C. Sarver, D. Silverman, & R. Weatherstone (Eds.), *Queer frontiers. millennial geographies, genders, and generations* (pp. 84–110). Madison: University of Wisconsin Press.

Holmes, M. (2004). Locating third sexes. *Transformations.* Retrieved February 17, 2010 from www.transformationsjournal.org/journal/issue_08/article_03.shtml.

Irigaray, L. (1985). *Speculum of the other woman.* Ithaca, New York: Cornell University Press.

Irigaray, L., & Gynn, N. (1995). The question of the other. *Yale French Studies, 87,* 7–19.

Kessler, S. (1998). *Lessons from the intersexed.* London: Rutgers University Press.

Khanna, R. (2003). *Dark continents. Psychoanalysis and colonialism.* Durham, London: Duke University Press.

Klöppel, U. (2002). "Störfall" Hermaphroditismus und Trans-Formationen der Kategorie "Geschlecht." Überlegungen zur Analyse der medizinischen Diskussionen über Hermaphroditismus um 1900 mit Deleuze, Guattari und Foucault. *Potsdamer Studien zur Frauen- und Geschlechterforschung, 6*(2), 137–150.

Leopold, J. (1980). *Culture in comparative and evolutionary perspective: E.B. Tylor and the making of primitive culture.* Berlin: Reimer.

Lévy-Bruhl, L. (1975). *The notebooks on primitive mentality.* Oxford: Basil Blackwell.

Lyons, A., & Lyons, H. (2004). *Irregular connections. A history of anthropology and sexuality.* Lincoln, London: University of Nebraska Press.

Malthus, T. (1798). *An essay on the principle of population, as it affects the future improvement of society with remarks on the speculations of Mr. Godwin, M. Condorcet, and other writers.* London: Johnson.

Markowitz, S. (2001). Pelvic politics: Sexual dimorphism and racial difference. *Signs: Journal of Women in Culture and Society, 26*(2), 389–414.

Mayne, X. (1908). *The intersexes. A history of similsexualism as a problem in social life.* Paris: Privately printed.

McClintock, A. (1995). *Imperial leather. Race, gender and sexuality in the colonial contest.* New York, London: Routledge.

Mitchell, K. (1997). Different diasporas and the hype of hybridity. *Environment and Planning, 15*(5), 533–553.

Neu, J. (Ed.). (1991). *The Cambridge Companion to Freud.* Cambridge: Cambridge University Press.

Oudshoorn, N. (2001). On bodies, technologies, and feminisms. In A. Creager, E. Lunbeck, & L. L. Schiebinger, (Eds.), *Feminisms in twentieth-century. Science, technology, and medicine* (pp. 199–213). Chicago: University of Chicago Press.

Probyn, E. (1993). *Sexing the self.* London: Routledge.

Said, E. (1978). *Orientalism.* New York: Vintage Books.

Schiebinger, L. (1989). *The mind has no sex? Women in the origins of modern science.* Cambridge: Harvard University Press.

Schiebinger, L. (1993). *Nature's body: Gender in the making of modern science.* Boston: Beacon Press.

Seshadri-Crooks, K. (1994). The primitive as analyst: Postcolonial feminism's access to psychoanalysis. *Cultural Critique, 28,* 175–218.

Sharp, L. A. (2000). The commodification of the body and its parts. *Annual Review of Anthropology, 29,* 287–328.

Somerville, S. (1994). Scientific racism and the emergence of the homosexual body. *Journal of the History of Sexuality, 5*(2), 243–266.

Somerville, S. (2000). *Queering the color line: race and the invention of homosexuality in American culture.* Durham: Duke University Press.

Spencer, H. (1897). *The principles of sociology.* New York: D. Appleton and Co.

Spencer, J. (2001). Ethnography after postmodernism. In P. Atkinson, A. Coffey, S. Delamont, & J. Lofland (Eds.), *Handbook of ethnography* (pp. 443–452). London: Sage.

Spivak, G. C. (1989). Who claims alterity? In B. Kruger and P. Mariani (Eds.), *Remaking history: Discussions in contemporary culture* (pp. 269–292). Seattle: Bay Press.

Spivak, G. C. (1993). Echo. *New Literary History, 24*(1), 17–43.

Stauder, J. (1993). The "relevance" of anthropology to colonialism and imperialism. In S. Harding (Ed.), *The "racial" economy of science. Toward a democratic future* (pp. 408–432). Bloomington, Indianapolis: Indiana University Press.

Stepan, N. L. (1986). Race and gender: The role of analogy in science. *Isis, 77* (2), 261–277.

Stepan, N. L. (1993). Appropriating the idioms of science: The rejection of scientific racism. In S. Harding (Ed.), *The "racial" economy of science. Toward a democratic future* (pp. 170–200). Bloomington, Indianapolis: Indiana University Press.

Stepan, N. L. (1996). Race and gender: The role of analogy in science. In E. Fox Keller and H. E. Longino (Eds.), *Feminism and science* (pp. 121–136). Oxford, New York: Oxford University Press.

Stoler, A. (1996). *Race and the education of desire: Foucault's history of sexuality and the colonial order of things.* Durham: Duke University Press.

Stoller, R. (1968). *Sex and gender. On the development of masculinity and femininity.* New York: Science House.

Street, B. (1990). Orientalist discourses in the anthropology of Iran, Afghanistan and Pakistan. In R. Fardon (Ed.), *Localizing strategies. Regional traditions and ethnographic writing* (pp. 240–259). Edinburgh: Scottish Academic Press.

Tylor, E. B. (1958 [1871]). *Primitive culture: Searches into the development of mythology, philosophy, religion, language, art, and custom.* New York: Harper & Row.

Tylor, E. B. (1964 [1865]). *Researches into the early history of mankind and the development of civilization.* Chicago: University of Chicago Press.

Traub, V. (1999). The psychomorphology of the clitoris. In S. Hesse-Biber, C. Gilmartin, & R. Lydenberg (Eds.), *Feminist approaches to theory and methodology: an interdisciplinary reader* (pp. 301–329). Oxford: Oxford University Press.

Weston, K. (1993). Lesbian/gay studies in the house of anthropology. *Annual Review of Anthropology, 22*(1), 339–367.

Werbner, P., & Modood, T. (Eds.). (1997). *Debating cultural hybridity: Multi-cultural identities and the politics of anti-racism.* London: Zed Books.

Young, R. (1995). *Colonial desire. Hybridity in culture, theory and race.* London: Routledge.

Conclusion

This book traces the constant reiteration of maleness/masculinity and femaleness/femininity as exclusive positions in the Western heterorelational sex-gender-sexuality system in cross-cultural intersexualization. On the basis and continuing attempt of the surgical and hormonal construction of two exclusive body morphologies (male and female), the gender-concept was developed. This gender-concept is based on the distinction between sex and gender and the pathologization of any possible incongruence between the two. It was not my intention to present a prescription of how to approach intersexualization, I simply attempted to show how the discursive underpinnings of intersexualization intersect with processes of racialization in neo_colonial settings. People might have different objectives they want to consider in political action aimed at the discrimination of non-normative gender identities, such as the possibility of surgery (as in the case of transsexualization) or the option to not have surgery (as in the case of intersexualized people).

Intersex activists continue reclaiming their history and challenge being called a phenomenon. In this volume I intended to demonstrate that the processes of intersexualization are the phenomenon. In order to do this, I treated the 'experts' in intersexualization as my case studies and interrogated how these experts make their results plausible, which concepts they draw on and how they position themselves in this endeavor. I thereby focused on the development of the gender-concept in the 1950s, as well as outlining how this concept embodies the epistemology of the West since the 1950s until now. Moreover, I zoomed in on an anthropological researcher's neo_colonialist moves to explain other cultures by superimposing this epistemology. I developed the concept of cross-cultural intersexualization in the two theoretical spaces of the Clinic and the Colony. My proposal was that in these two spaces the phantasm of intersexuality haunts the researchers who pursue the quest to find a viable distinction between men and women and the hermaphrodite/intersexualized.

In line with Foucault, I argued that knowledge is not made for understanding but for cutting. I played with this figure on five levels. First of all, the knowledge that is produced in intersexualization is made for the cut in the hierarchical dichotomy between the sexes_genders. This cut works on the premise that there are men and women who behave in a certain way and inhabit certain roles in

society, which subsequently require securing to protect the status quo of the hierarchical split. The second cut that is made is in the distinction between sex and
gender, which is the cut between the morphological and the social or psychological. Researchers in intersexualization, as I have demonstrated, use this cut to determine the normality of congruence between the two; women are supposed to have a
specific morphology and men another. If this is not the case or if this congruence
is somehow at stake then the mode of cutting enters the sphere—the separation
between the normal and the abnormal. This is based on the congruence or incongruence of sex and gender. If the sex of a body cannot be identified properly or if
the identity of the individual does not align to the body from which it should
follow, it is determined abnormal. The fourth layer then enters the realm of intersexualization. This fourth level refers to the surgical practices in intersex management, which use the scalpel to guarantee the first level of the distinction between
the sexes_genders and the second of aligning the morphology and the social or
psychological. The cut in intersexualization is represented by the actual cut into
the flesh of the body. Most of the time this cut features as clitorectomy, the amputation of 'phallic flesh' in women, which testifies to the masculinist and sexist
underpinnings in intersexualization. The fifth level is represented by the fifth other
that is being cut from the heterosexual, white, Euro- and US-Americancentric
notion of the norm—the non-hybrid, single-sexed, and single-cultured norm. In a
globalized and neo_colonial age, this other-than-the-binary-sexed-body and other-
than-the-Western-body, represented by the cross-cultural intersexualization of
Caster Semenya, the Turmin-Man Sakulambei of Papua New Guinea, and the
guevedoche of the Domincan Republic are cut out as presumably ambiguous and
insufficiently developed in the logics of cross-cultural intersexualization that
depend on the notion of development towards.

However, I acknowledge that through my position as a researcher who com-
poses a genealogical account, just as the one presented in this book, I also cut.
Yet, I hope that I contribute to the growing understanding of how intersexualization works. In this volume I decided to use certain moments in the history of
knowledge production to interrogate a specific body of knowledge. I made a
decision between those knowledge productions which I considered as important
and those which I do not see as equally influential. I also cut out the specific time
frame of the period after the Second World War and the period of the late 1970s
to early 1980s in which the theories of the 1950s were recycled in research on
other cultures. Therefore, I also made a geographical incision, namely by focusing on researchers in the USA and their travels to Papua New Guinea and the
Dominican Republic. As a researcher I have chosen particular instances in the
processes of intersexualization to show how they work at exactly these points in
time and space. I slice time and space and according to importance and relevance for my argument. My contribution to understanding, however, is how these
cuts are made and how these lines are drawn in the first four areas in the particular historical instances I analyzed.

It is important to state that I consciously chose to not explain what so-called
'intersex conditions' or Disorders of Sexual Development are. I deliberately

avoided repeating the cut between the normal and the abnormal. With this book I tried to avoid being complicit in the repeated articulation of diagnoses, which medicalize and pathologize bodies not in medical need of surgery. Only in a few so-called intersex conditions are surgeries medically necessary. (Some cases of CAH have severe salt loss which can be deadly.) They require treatment; however, this does neither require the diagnosis of intersexuality nor DSD, nor surgery. Most of the surgeries, which were and are still performed, serve to consolidate heterorelational sexual difference and clear-cut sexed_gender identities. The fact that surgeries are performed on some bodies to align them to sexual dimorphism only proves that the cut in sexual dimorphism is a phantasm.

This fantasy, however, is constantly reaffirmed in the processes of cross-cultural intersexualization. Multiple theories and references are called upon to establish the norm of the two sexes_genders as the natural expression of sexual difference. Medical and psychological experts apply a framework of the normal, which I have called the 'operation theater' and make it seem natural. This 'operation theater' is prepped with concepts such as gender role and gender identity, or by actual surgical tools such as the scalpel.

I intended to show that intersexualization since the 1950s in John Money's and Robert Stoller's research is based on the distinction between sex and gender. In their intersexualizing research the gender-concept was invented in order to determine what is normal and what is abnormal for a body to behave and to feel. The quest for a scientifically verifiable distinction between men and women is pursued in the theories which invent categories such as gender identity and gender role; yet, this always comes down to biological determinism, even though gender as the verification tool is called upon as in the case of Caster Semenya. The decision in the social is first made on the morphological level; then, the social separation is reaffirmed on the basis of sex verification. Semenya could only continue to run in the women's camp because her body testifies to the parameters of femaleness.

Of course, at different moments during the last 50 decades Money and Stoller had developed their theories while shifting their emphasis. Intersexuality, as Morgan Holmes states, is a shifting phantasm in medical discourse which is constantly rearticulated, rephrased, and modified. However, I focused on particular moments in the history of intersexualization, which were crucial for the treatment paradigm to be established and the gender-concept to be verified by the invention of the third sex and third gender. This was later adapted by Herdt to explain the culture of the Sambia in Papua New Guinea.

I focused on the research by Money and Joan and John Hampson in which they created a field of expertise that constantly feeds itself by creating the problem of intersexuality. Their sample of 'patients' showed no significance in regards to psychopathology; there were adult women who had 'phallic flesh' and showed no psychotic symptoms. However, Money and the Hampsons disregarded their results and drew the conclusion that surgical intervention on infants is necessary to secure a stable gender role. That Money et al. were able to do this reflects developments in the waxing privileges of psycho-medical experts since the scientific revolution; their treatment recommendations are still in place. Notably, these recommendations had atrocious effects on now living adults who were diagnosed as psycho-sexual emergencies and subjected to surgical intervention as infants. Some of these adults speak out now and challenge the mutilating practices in Western hospitals. Recently, far reaching discussions also inside the medical establishment have begun and researchers review their treatment paradigms. However, the new nomenclature of Disorders of Sexual Development is emblematic for the knowledge production which is not made for an understanding of the multiplicity of bodies. Rather, it advocates the continuous cut between the normal and the abnormal. Endocrinology, genetics, and neuropsychology are only the tip of the gender iceberg, as Anne Fausto-Sterling calls it. With these conceptual tools in hand, scientific experts proceed through bodies, penetrate tissue, and colonize the psyche in order to find verifiable distinctions, in accordance with the requirement and circumstances of the sociopolitical atmospheres. In the second part of the introduction to this volume on *From Myth to Medicalization* I described the dawn of the heroic age of the scientist as a colonial agenda, which diagnoses and gives names, categorizes, and classifies human beings in order to assign them a place in the orders of nature. The phenomenon of intersexualization produces the notion of an 'experiment in nature' and then denotes it as 'abnormal.' Intersexualization then erases the non-normative through surgery to reaffirm the norm. The cut is the normative cut of a heterorelational organization of the sexes_genders.

Just as intersex movements gained power and challenged medical expertise, the term intersex has been replaced by the term Disorders of Sexual Development in 2005. DSD is a term which silences the fact that surgery produces embodiment. DSD veils the effects intersexualization has on the embodiment of intersexualized infants and limits the space of possibility for embodiment without surgical experience. Moreover, it reinstalls the notion of development as being a basic feature of (cross-cultural) intersexualization. To reinstall the notion of disorders in sexual development is to call upon these discourses and to pathologize that which disrupts the heterorelational organization of the sexes_genders. In the part on the Clinic, I demonstrated how John Money and Robert Stoller, in their psychological and psychoanalytical theories of gender role and gender identity, appropriated concepts such as innate bisexuality, sexual identification, and psychosexual maturity. These concepts are modified in their intersexualizing theories in relation to the postulated necessity of a development towards a heterosexual monosexuality. Intersexualization has thus a specific place in the order of theories on 'normal gender identity development' and 'normal sexual differentiation' because the category of intersexuality is the foundation for reformulations of the norms of maleness and femaleness as well as femininity and masculinity. Stoller invented the category of core gender identity to argue for an innate biological force (sex) that causes the feeling of being female or male. He later added a hermaphroditic core gender identity to these two and installed a third available position on the basis of biological determinism. For this third identity he created a category that according to him not really belongs to the human race. He advocated surgery to normalize individuals whom he assigned this category. I identified this as the rhetorical mode of first naming the exception to the norm to then normalize it; it is the production of a hierarchy of the normal in the so-called natural development of the human towards sexual dimorphism.

Yet, when intersexualization enters the space of the Colony, the notion of development produces a second rhetorical mode with regards to the Western orders of nature. In the space of the Colony the designation of hermaphroditism/intersexuality as a 'natural experiment' in a more 'natural setting' is emblematic for this process. The Clinic of the Global North exports its modes of dissecting, displaying, diagnosing, and exhibiting to cultures of the Global South.

As I have elaborated in the introduction, Caster Semenya's presence in the media is not so different from the spectacle that Sarah Baartman was subjected to in the nineteenth century; Semenya has been othered by the representation of her body as not properly sexed and not properly gendered according to the standards of white, female, bourgeois, and civilized standards of Western femininity. Additionally, she has been othered through silencing. She has been denied a subject status from which she can claim authority over the description of her body and her identity. I argued in this book that silencing and invisibilization is a dominant feature in cross-cultural intersexualization. In the case study on clinical ethnography by Herdt and Stoller I analyzed how Sakulambei is silenced and invisibilized. Herdt's and Stoller's paternalistic attitude features in their behavior towards Sakulambei and in their claim to be able to extract the secret of sex of Sakulambei's body. Here the medical Western, hegemonic gaze of the Clinic becomes a gaze which produces not just the body of the other (the intersexualized) but also produces the culture of the other, which is less sophisticated or civilized than Western culture. Subsequently, it can, therefore, host the hermaphrodite as a part of its culture. Through its being called polymorphous perverse–and therefore permissive towards sexual variety–the other culture is, as whole, bound to a naïve and childlike status.

By analyzing Stoller's work in the Clinic and Herdt's adaptation of it in the Colony, I presented the ways in which the construction of a 'third' in intersexualization functions and serves the purpose of consolidating a hermaphroditic gender identity on the grounds of a hermaphroditic body. This consolidates the assumption that a male body causes a male gender identity and a female body causes female identity, which is stable throughout life. In this framework, the third gender becomes the cultural yet necessary expression of a third sex. By using the concept of the polymorphous perverse to explain a postulated three-sex system in Sambian culture, this specific system is exoticized and constructed as less developed, less civilized; the two-sex one becomes newly institutionalized as hegemonic in the space of the Colony. Even though I charted the polymorphous perverse as a promising concept in Freud's work, which could have been taken up by Money and Stoller instead of the concept of innate biology, I do see a problem in using it for the description of the organization of another culture as a whole. By championing the polymorphous perverse as a possibility to think embodiment, desire, and subjectivity differently in the West I intended to indicate that the notion of bifurcated identification process in an oedipal setting reaffirms civilization as necessarily based on two heterorelational mono-sexualities. Yet, using the polymorphous perverse to describe the permissiveness to the multiplicity of embodiment of other cultures, as I demonstrated in the case in Herdt's work, is to imply the framework of civilization. This subsequently positions the other culture as childlike and less developed in this psycho-analytical framework. I do not doubt Herdt's intention to defy the rigid sex_gender system of the West by referring to the Sambia who are able to accommodate more than two sexes_genders. Yet, the use of this metaphor of the polymorphous perverse is not sufficiently explained; civilization as the celebra-tion of binary sexual difference is not tackled but rather reiterated and rein-stalled. Moreover, to use the sexuality of the other always implies otherness on the basis of sexuality. Normalizing processes that occur in the West are based on references to the other. The distinction between male/masculine and female/fem-inine as a product of these normalizing processes is in its production not just based on the referent of the 'sex(ualized)_gendered other' but also based on the referent of the 'racialized/ethnicized other.'

In the part on the Colony I interrogated the underlying evolutionary dis-courses which ground the othering process fundamental to cross-cultural inter-sexualization. I hereby focused on the discourses of anthropology, psychoanalysis, and sexology as having emerged in a colonial and imperialist time which was dominated by (social) Darwinist evolutionary thinking. The orders of nature from this time are clearly present in the revived notion of the fifth other, the polymorphous perverse sexed savage in cross-cultural intersexu-alization. These orders are reinstalled when Herdt argues that the Sambia, who are thereby represented as immature and childlike as a culture, can accommodate a third gender that is based on a third sex. The civilized and sophisticated West produces the other and their 'other sexuality,' namely that of arrested develop-ment of the intersexualized body, in the arrested development of the other culture. In the Colony I intended to show that the phallocratic and ethnicized/racialized organization of bodies along the hierarchical coordinates of 'white' maleness/masculinity and femaleness/femininity and their locatedness in geo-political arrangements of bodies and identities has not ceased to be employed in the interpretation and representation of the 'other body' in the 'other culture.' Intersexuality in the other is represented as a mythical 'phenomenon.' The story of the other is silenced by the imposition of the story of the West and its orders of nature.

However, in the excursus I very briefly looked into the reclaiming of stories by the Two-Spirit movements of Native Americans. By tracing the application of the anthropological term *berdache* I demonstrated that multiplicities of cul-tural organizations in Native American cultures were also subsumed under one umbrella term. Anthropologists then used to describe what they cannot grasp but nevertheless wanted to categorize and make intelligible in the framework of a Western bio-medical, sexological, and sometimes also psychoanalytical explanation system of sexual dimorphism and the organization of binary genders. The Two-Spirit movements re-collect their stories and their histories and re-assemble them in a post-colonial space. These movements emphasize and combine differences in a third space in which various voices reclaim and adapt various stories. They compose a future that incorporates their histories, yet produce something new. Their assembling and reclaiming of different narratives seems to me a promising move which could also help open up the dichotomous Western sex-gender-sexuality-system and make it obsolete. I want to end this book on intersexualiza-tion with a plea against epistemic and material-discursive violence i.e., the cutting scalpel and *for* the multiplicity of narratives, identities, and embodiments.

Index

Taylor & Francis eBooks

Helping you to choose the right eBooks for your Library

Add Routledge titles to your library's digital collection today. Taylor and Francis ebooks contains over 50,000 titles in the Humanities, Social Sciences, Behavioural Sciences, Built Environment and Law.

Choose from a range of subject packages or create your own!

Benefits for you

» Free MARC records
» COUNTER-compliant usage statistics
» Flexible purchase and pricing options
» All titles DRM-free.

REQUEST YOUR FREE INSTITUTIONAL TRIAL TODAY

Free Trials Available
We offer free trials to qualifying academic, corporate and government customers.

Benefits for your user

» Off-site, anytime access via Athens or referring URL
» Print or copy pages or chapters
» Full content search
» Bookmark, highlight and annotate text
» Access to thousands of pages of quality research at the click of a button.

eCollections – Choose from over 30 subject eCollections, including:

Archaeology	Language Learning
Architecture	Law
Asian Studies	Literature
Business & Management	Media & Communication
Classical Studies	Middle East Studies
Construction	Music
Creative & Media Arts	Philosophy
Criminology & Criminal Justice	Planning
Economics	Politics
Education	Psychology & Mental Health
Energy	Religion
Engineering	Security
English Language & Linguistics	Social Work
Environment & Sustainability	Sociology
Geography	Sport
Health Studies	Theatre & Performance
History	Tourism, Hospitality & Events

For more information, pricing enquiries or to order a free trial, please contact your local sales team:
www.tandfebooks.com/page/sales

Routledge
Taylor & Francis Group

The home of
Routledge books

www.tandfebooks.com